Feedlot Ruminant Nutrition and Metabolism

Feedlot Ruminant Nutrition and Metabolism

Guest Editors

Miguel Henrique De Almeida Santana
Nara Regina Brandão Cônsolo
José Bento Sterman Ferraz

MDPI

Basel • Beijing • Wuhan • Barcelona • Belgrade • Novi Sad • Cluj • Manchester

Guest Editors
Miguel Henrique De Almeida Santana
Department of Animal Sciences
University of Sao Paulo
Pirassununga
Brazil

Nara Regina Brandão Cônsolo
Department of Animal Nutrition and Production
University of São Paulo (USP)
Pirassununga
Brazil

José Bento Sterman Ferraz
Department of Veterinary Medicine
University of Sao Paulo
Pirassununga
Brazil

Editorial Office
MDPI AG
Grosspeteranlage 5
4052 Basel, Switzerland

This is a reprint of the Special Issue, published open access by the journal *Metabolites* (ISSN 2218-1989), freely accessible at: https://www.mdpi.com/journal/metabolites/special_issues/O57C5G9DM0.

For citation purposes, cite each article independently as indicated on the article page online and as indicated below:

Lastname, A.A.; Lastname, B.B. Article Title. *Journal Name* **Year**, *Volume Number*, Page Range.

ISBN 978-3-7258-6880-3 (Hbk)
ISBN 978-3-7258-6881-0 (PDF)
https://doi.org/10.3390/books978-3-7258-6881-0

Contents

About the Editors

Miguel Henrique de Almeida Santana

Miguel Henrique de Almeida Santana is an Associate Professor of Animal Science at the School of Animal Science and Food Engineering, University of São Paulo, where he conducts research and teaching in beef cattle production systems, nutrigenomics, and animal genomic science. He earned his Bachelor's degree in Animal Science (2008) and a Master's degree in Veterinary Sciences (2010) from the Federal University of Paraná, followed by a Ph.D. in Animal Science (2013) and a postdoctoral qualification (2018) from the University of São Paulo. He also holds the academic title of Livre-Docente awarded in 2022, reflecting advanced scholarly achievement in his field. His research program focuses on the genetic and genomic determinants of feed efficiency, nutrient–gene interactions, fetal programming effects in beef cattle, and the application of systems biology to enhance production outcomes. Professor Santana has been involved in international collaboration, including a visiting research appointment at the University of Copenhagen (2014–2015), and he coordinates and participates in multiple funded research projects on nutrigenomics and holistic approaches to bovine production. In his role as educator, he teaches undergraduate and graduate courses in animal genomics, nutrigenomics applied to production, and bovine science, and supervises research across multiple levels, including doctoral, master's, and undergraduate projects that explore the efficiency traits, metabolomics, and molecular biology of livestock. His scientific contributions have been recognized through awards such as the "Nicolau Athanassof" Prize for the best master's thesis in Zootecnia, as well as distinctions for presentations at scientific symposia, underscoring his impact on the field of animal science in Brazil.

Nara Regina Brandão Cônsolo

Nara Regina Brandão Cônsolo is a Professor at the Department of Nutrition and Animal Production at the School of Veterinary Medicine and Animal Science, University of São Paulo (FMVZ-USP), where she leads research and instruction in beef cattle production, nutrition, metabolomics applied to animal production, and meat science. She completed her Bachelor's degree in Animal Science at Universidade Estadual Paulista Júlio de Mesquita Filho in 2009, after which she obtained a Master's degree (2011) and a Doctorate (2015) in Nutrition and Animal Production at FMVZ-USP, both supported by FAPESP scholarships. She subsequently pursued postdoctoral research at the Department of Animal Science (FZEA-USP, Brazil), AgResearch Grasslands (New Zealand), and Universidade Federal da Grande Dourados (Brazil), with additional international research experiences at Texas Tech University and Canadian institutions including Dalhousie University and the University of Guelph. Her research program integrates advanced metabolomic and genomic approaches to understand and improve beef cattle performance, carcass and meat quality traits, and metabolic phenotypes, exemplified by projects such as a FAPESP-funded initiative on metabolomic strategies to enhance Nellore beef tenderness through metabotype and biomarker discovery. Professor Cônsolo has contributed to the field through mentorship of graduate and undergraduate researchers, presentations at international scientific meetings such as the International Congress of Meat Science and Technology, ad hoc review activities for scientific conferences, and the development of multidisciplinary collaborations that bridge nutritional biology, multi-omics data analysis, and applied livestock science, thereby advancing both fundamental understanding and practical applications in animal production systems.

José Bento Sterman Ferraz

José Bento Sterman Ferraz is a Full Professor at the Department of Veterinary Medicine, Group of Animal Breeding and Biotechnology, at the School of Animal Science and Food Engineering, University of São Paulo (FZEA-USP), where he has served at senior academic and research roles since 1997, and holds CNPq Researcher 1-A status. He graduated in Veterinary Medicine from the USP in 1977 and subsequently completed a Master's degree in Genetics (1980) and a Ph.D. in Genetics (1981) at the same institution, establishing a lifelong specialization in quantitative genetics and animal breeding. Professor Ferraz's research program focuses on genetic evaluation and selection methodologies for beef cattle and other livestock, including the validation of molecular markers, genome-wide association studies, genomic selection, and genetic parameter estimation, with applications that have shaped sire evaluation practices and breeding strategies in Brazil. He leads and supervises research on genetic improvement, has coordinated numerous research teams, and his scientific impact is reflected in citation metrics including an h-index of 36 in Scopus and 49 in Google Scholar, and an i10 index of 200, underscoring sustained influence in animal genetics and breeding. Professor Ferraz has also made substantial editorial contributions as Genetics, Animal Genetic Resources and Breeding Section Editor for the journal *Livestock Science* (Elsevier) since 2010 and was Editor of the Special Issue "Genomics Applied to Livestock Production" presented at the 10th World Congress on Genetics Applied to Livestock Production. In addition to institutional leadership roles such as Coordinator of Faculty Evaluation in Animal and Food Sciences at USP, he has held visiting professorships, including at the University of Lisbon, and received the Prof. José Rodolpho Torres Award from the Brazilian Society of Animal Breeding in recognition of his distinguished career in the genetic improvement of livestock.

metabolites

Article

Metabolomics Changes in Meat and Subcutaneous Fat of Male Cattle Submitted to Fetal Programming

Arícia Christofaro Fernandes [1,*], Guilherme Henrique Gebim Polizel [1], Roberta Cavalcante Cracco [1], Fernando Augusto Correia Queiroz Cançado [1], Geovana Camila Baldin [1], Mirele Daiana Poleti [2], José Bento Sterman Ferraz [2] and Miguel Henrique de Almeida Santana [1]

1 Department of Animal Science, College of Animal Science and Food Engineering, University of São Paulo (USP), Av. Duque de Caxias Norte, 225, Pirassununga 13635-900, SP, Brazil; guilherme.polizel@usp.br (G.H.G.P.); mhasantana@usp.br (M.H.d.A.S.)
2 Department of Veterinary Medicine, Faculty of Animal Science and Food Engineering, University of São Paulo (USP), Av. Duque de Caxias Norte, 225, Pirassununga 13635-900, SP, Brazil; mirelep@usp.br (M.D.P.); jbferraz@usp.br (J.B.S.F.)
* Correspondence: aricia.fernandes@usp.br or ariciacf@gmail.com; Tel.: +55-119421427

Abstract: This study investigated changes in meat and subcutaneous fat metabolomes and possible metabolic pathways related to prenatal nutrition in beef cattle. For this purpose, 18 Nellore bulls were used for meat sampling and 15 for fat sampling. The nutritional treatments during the gestation were: NP—not programmed or control, without protein-energy supplementation; PP—partially programmed, with protein-energy supplementation (0.3% of body weight (BW)) only in the final third of pregnancy; and FP—full programming, with protein-energy supplementation (0.3% of BW) during the entire pregnancy. The meat and fat samples were collected individually 24 h after slaughter, and the metabolites were extracted using a combination of chemical reagents and mechanical processes and subsequently quantified using liquid chromatography or flow injection coupled to mass spectrometry. The data obtained were submitted to principal component analysis (PCA), analysis of variance (ANOVA), and functional enrichment analysis, with a significance level of 5%. The PCA showed an overlap between the treatments for both meat and fat. In meat, 25 metabolites were statistically different between treatments ($p \leq 0.05$), belonging to four classes (glycerophospholipids, amino acids, sphingolipids, and biogenic amine). In fat, 10 significant metabolites ($p \leq 0.05$) were obtained in two classes (phosphatidylcholine and lysophosphatidylcholine). The functional enrichment analysis showed alterations in the aminoacyl-tRNA pathway in meat ($p = 0.030$); however, there was no pathway enriched for fat. Fetal programming influenced the meat and fat metabolomes and the aminoacyl-tRNA metabolic pathway, which is an important candidate for the biological process linked to meat quality and related to fetal programming in beef cattle.

Keywords: meat quality; metabolites; muscular and adipose development; pregnancy; prenatal supplementation

1. Introduction

Nutrition is one of the environmental factors that can most influence the phenotype, and that producers are able to handle with some ease. However, knowledge on the internal physiological factors that control the production, well-being, and quality of its products has not been fully elucidated [1]. There are several gaps in understanding the mechanisms for obtaining certain phenotypes in beef cattle. Meat production, for example, in its most literal and basic sense constitutes the formation and growth of the animal's muscular and adipose tissue [2]. For this, the individual needs to have substrate for this formation, which comes from nutrition and tissue metabolism. These tissues originate during the animal's pregnancy [3]. Therefore, the supply of nutrients from the cow during pregnancy is responsible for ensuring adequate nutritional support for the development of these tissues in the progeny [4].

Carcass traits and meat quality are influenced by several metabolic pathways, based on the sum of interactions between genotype, environment and the interaction of both [5]. The metabolome is defined as the complete set of small molecules (metabolites) that participate in metabolism [6]. Metabolites are the result of the complex interactions that occur between the genome and the environment, and the science that understands how this mechanism occurs is metabolomics [7]. Therefore, the advancement of molecular biology and the advent of metabolomics could contribute to animal research and to the understanding of complex biological systems, such as the interaction of metabolites and pathways resulting in the expression of phenotypic characteristics [8]. The application of metabolomics comes to identify and quantify the characteristic metabolome that serves as a substrate for the formation of tissues of interest, allowing gains in meat quality and quantity in herds [9].

Studies using metabolomics in meat have emerged in recent years [10,11], as well as studies of the metabolome between meat produced under different animal feeding conditions [12,13]. In the study by Zhang et al. [14], it is possible to find information on the relationship between metabolites and aspects of meat quality such as color, pH, tenderness, etc. The work of Antonelo et al. [15] uses metabolomics as a tool for identifying variations in meat tenderness. In the research by Zuo et al. [16], metabolomics is approached as a way to explain at the metabolic level how a characteristic of meat (water retention capacity) is different according to maturation times. And a review presented by Bischof et al. [17] compiles several other studies that evaluate how metabolomic changes occur in meat during the transformation of muscle into meat after slaughter for long periods. These are all recent works that show the importance of meat metabolomics.

Knowledge of the metabolomic profile in meat/fat helps to compose the observed phenotype, based on the distinction of metabolites, so the quantity and type of metabolite can determine the physiological characteristics of the muscle and the quality characteristics of the meat [18]. The observation of changes in progeny resulting from interventions during pregnancy is known as fetal programming; several studies have correlated fetal programming with the development of muscle and adipose tissues [3,19,20]. Some studies have already tested different feeding strategies and characterized the metabolome of animals destined for meat production [12,21–25]. However, few studies have investigated the consequences of prenatal nutrition on the metabolism of their offspring [26–29].

This study aims to evaluate the meat and subcutaneous fat metabolome of young bulls submitted to fetal programming. Our hypothesis was that protein-energy supplementation at different periods of gestation or its absence in bovine cows alters the muscle and adipose metabolome of male progeny.

2. Methodology

2.1. Declaration of Ethics for the Use of Animals in Experimentation

All experimentation protocols involving these animals were previously approved by the Research Ethics Committee of the Faculty of Animal Science and Food Engineering of the University of São Paulo (under protocol CEUA n° 1843241117, on 10 March 2018), in accordance with the guidelines of the National Council for the Control of Animal Experimentation (CONCEA).

2.2. Experimental Design

A sample of 126 Nellore females was submitted to fixed-time artificial insemination using semen from four bulls. The females were multiparous and were 3–8 years old. After confirmation of pregnancy, the cows were selected and randomly assigned to treatments according to age, body weight, and body condition score, measured at the time of insemination in order to keep the batches as homogeneous as possible. All cows remained in paddocks with pasture of *Brachiaria brizantha* cv. Marandu throughout the experiment. All females received mineral supplementation (0.03% average body weight of each batch) and water ad libitum throughout pregnancy.

Each cow was considered an experimental unit and was assigned to a completely randomized design with three treatments, being prenatal nutrition strategies (i.e., fetal programming), which consisted of providing a protein-energy supplementation during pregnancy. The treatments were: not programmed or control (NP), without protein-energy supplementation; partially programmed (PP), with supplementation only in the final third of pregnancy; and full programming (FP), with protein-energy supplementation throughout the pregnancy until calving. Both the PP and FP groups received a daily supplement corresponding to 0.3% of the average body weight of the cows until calving, in accordance with the National Research Council's (NRC, 2000) [30] nutritional, maintenance, and gain recommendations for cows during mid- to late pregnancy. The ingredients and nutrients in the supplement offered to pregnant cows and the nutrients in pastures can be found in the methodology of Schalch Junior et al. [26].

After calving, all animals received the same sanitary, nutritional, management, and experimental collection conditions. All male progeny (63 animals, 21 per treatment) spent 8 months with the mother until weaning and another 11 months in the rearing phase on pasture and supplementation in the trough, followed by an average of 106 days of a finishing phase in the feedlot. At calving, all animals received conventional care (weighing, navel healing, deworming, and identification tattoo) and were kept under the same pasture system (rotated in *Brachiaria brizantha* cv. Marandu) staying with their dams until weaning. Afterward, they passed to the rearing phase, under the same forage species but in two paddocks, and additionally now receiving an energy supplementation of 0.3% of the average weight of the lot during the dry period and protein supplementation of 0.1% of the average weight of the lot in water during the rearing period until the animals entered the feedlot.

The animals were divided into two collective pens according to the entry weight upon arriving at the feedlot facilities and were slaughtered after the finishing period. The slaughter was carried out in accordance with humanitarian procedures, as required by Brazilian law [31,32]. The slaughter was carried out in compliance with humane procedures, as required by Brazilian law. The slaughter procedure occurred in three groups of 21 animals each (7 animals per treatment) and with the slaughter of one group per week, but all 21 animals of each group were slaughtered on the same day. The selection criterion of the animals for each slaughter group was the subcutaneous fat thickness (SFT) obtained in the last carcass ultrasound on the penultimate day of the finishing phase (pre-slaughter). Based on these data, slaughter began with the group of animals with the highest SFT in each treatment. After slaughter, the carcasses were placed in refrigeration chambers with a temperature between −4 and 0 °C for 24 h.

The males were subjected to evaluations involving body weight measurements from birth to slaughter. Only during weaning (8 months) were the groups differentiated. During this period, the PP and FP groups demonstrated a greater body weight compared to the NP group (control), although this distinction was not manifested at any other time, as reported by Cracco et al. [33]. The measurements of the groups' weights over the period can also be consulted in the cited work.

The animals used for tissue metabolomics were selected from a subsample, more specifically 18 males for meat metabolomics, 6 of each treatment, while 15 of the 18 animals (5 from each treatment) were selected for fat metabolomics. To ensure greater data homogeneity, the animals were selected based on the premise that they were offspring of the same father and that the mothers were born in the same year.

2.3. Tissue Collection

Samples of meat and adipose tissue were used to carry out the analyses. The meat tissue samples were obtained from the *Longissimus thoracis et lumborum* muscle, on the sectioned face between the 12th and 13th rib, and the subcutaneous fat from this same muscle and in the same position; both were obtained 24 h after slaughter.

In the deboning room, the sirloin steak was separated from the carcass and arranged for the seizure of biological material. Both tissues were collected at the time of deboning and handled as hygienically and quickly as possible. The samples were aliquoted and placed individually in liquid nitrogen in previously autoclaved aluminum foil packaging. The meat samples were removed in a similar position between the animals, at the central point of the cut and the fat layer, on the sectioned face between the 12th and 13th rib. Small samples were removed with tweezers and a scalpel, which were sanitized and free of cross-contamination, and then stored in an ultrafreezer at −80 °C until extraction.

2.4. Tissue Homogenization and Metabolite Extraction

According to the methodology proposed by Zukunft et al. [34], for skeletal muscle and fat tissues, there is a specific solvent and an indicated proportion. In the case of the solvent (same for both tissue), each type of tissue was homogenized in two different reagents: 10 mM phosphate buffer pH 7.5 at 25 °C and ethanol 85/15 (v/v)/10 mM mixture of phosphate buffer pH 7.5 (EtOH/PB). As for the proportions [mg of tissue to X µL of solvent, denoted by 1:X (w/v)], the proportion 1:3 was used for muscle and 1:6 for fat.

2.5. Preparation of Extraction Solvent

A mixture of the two reagents mentioned above and water was used to compose the total extraction solvent. Initially, we used solvent C composed of 10 mL of 0.1 M phosphate buffer +90 mL of water (HPLC). For the formation of solvent A, 15 mL of solvent C was added to 85 mL of ethanol (HPLC-grade), forming the specific solvent for the extraction of muscle and fat samples.

The samples were removed from the aluminum foil and weighed for homogenization, and 30 g was the tissue weight used for extraction. Then, the samples were placed in pre-cooled cryotubes (dry ice) containing ceramic spheres with a diameter of 1.4 mm (Precellys Homogenization Kit, CK14, PEQLAB Biotechnology, Erlangen, Germany). Subsequently, the appropriate solvent proportion of each tissue was added into the cryovials, with 30 g of meat/90 mL of solvent and 30 g of fat/180 mL of solvent.

After freezing at −20 °C, the extraction solvent was added to the cryotubes with the samples that were previously weighed. Then, the samples were homogenized in a Precellys®24 homogenizer equipped with an integrated cooling unit (PEQLAB Biotechnology, Erlangen, Germany) three times for 20 s at 5500 rpm with 30 s intervals to ensure constant temperatures during homogenization [34].

After, the samples were centrifuged at 10,000 rpm for 5 min at +4 °C to separate the metabolites. The supernatant generated from these procedures was pipetted into a new vial and stored at −80 °C. For the metabolomic analysis, 10 µL of the supernatant was used.

2.6. Targeted Metabolomics

The metabolomic analysis was carried out by the company Apex Science (Campinas, São Paulo, Brazil). The AbsoluteIDQ ®Kit p180 (Biocrates Life Sciences AG, Innsbruck, Austria) was the product used for the metabolite quantification. The kit comprised 188 metabolites, 21 of which were amino acids, 21 biogenic amines, 40 acylcarnitines (Cx:y), 14 lysophosphatidylcholines (lysoPC), 76 phosphatidylcholines (PC), and 15 sphingolipids (SMx:y), where "x" represents the number of carbons and "y" the double bonds of all chains. The amino acids and biogenic amines were derivatized using phenylisothiocyanate. These classes of metabolites were analyzed using liquid chromatography–tandem mass spectrometry (HPLC–MS/MS) using an AB Sciex 4000 QTRAP mass spectrometer (AB Sciex, Darmstadt, Germany) with electrospray ionization. The lysophosphatidylcholines, phosphatidylcholines, acylcarnitines, and hexose were analyzed using injection flow–tandem mass spectrometry (FIA–MS/MS).

To carry out the data analysis of metabolite quantification and quality assessment, the MetIDQ® v1.0 software (part of the AbsoluteIDQ® p180 kit) was used. The metabolite concentrations (measured in μM) were calculated using the internal standards. The enterprise Biocrates Life Sciences AG determined experimentally the metabolite-specific detection limits (LOD) of the assay.

The metabolite quantification methodology, kit validation, internal standards for the liquid chromatography method, and concentrations of reference analytes for quality control (QC1–3), carried out by Apex Science, were replicated from Zukunft et al. [34], being a protocol established by the commercial kit supplier and adapted for tissues.

2.7. Statistical Analysis

The data processing and univariate analysis (analysis of variance—ANOVA) of the metabolites were performed in the R software environment (version 4.1.2). Metabolites with more than 70% of samples below the LOD or with the same values across samples were removed from the dataset. The LOD values that remained in the metabolome after filtering were replaced by the mean of each variable. The model was implemented using the "LM" function in R.

The statistical model used for the metabolomic and phenotypic analyses of bulls was:

$$Yjk = \mu + \beta 1Ageb1 + Tratj + \varepsilon jk$$

where Yjk is the observed metabolite of the kth animal, recorded in the jth treatment; μ is a constant; β1 is the regression coefficient of the animal age covariate; Ageb1 is the observed value for the animal age of the kth animal; Tratj is the fixed effect of the jth treatment; and εjk is the residual random term. The residuals were tested for normality (Shapiro–Wilk test) and homoscedasticity (Levene test) and differences between treatments were considered significant when $p \leq 0.05$ according to the Tukey–Kramer test.

In addition, the concentration of metabolites was analyzed using the MetaboAnalyst 5.0 software and the data were auto-sized (centered on the mean and divided by the standard deviation of each variable) before analysis. We performed the principal component analysis (PCA) and the functional enrichment analysis of metabolic pathways for the significant metabolites in the meat and fat samples. The PCA was performed to evaluate the clustering between treatments (NP, PP, and FP). The enrichment analysis was performed to identify the most relevant biological processes associated with the differentially expressed metabolites (identified in the univariate analysis) based on the Kyoto Encyclopedia of Genes and Genomes database (KEGG Pathway) and PubChem (Open Chemical Database on National Institutes of Health (NIH)). These databases allow you to find different biological pathways (e.g., energy, lipid, and amino acid metabolism) related to input (differential expressed metabolites) and molecule functionality. Biological processes with a p value ≤ 0.05 were considered significant.

3. Results

3.1. Principal Component Analysis (PCA)

The distribution of all data analyzed showed an overlap between all groups and there was no clustering between the treatments. The overlap indicates similarity between the treatments. The two principal components together for meat explain 54.9% of the total variance (PC1 = 36.3%; PC2 = 18.6%), whereas for fat, 52.0% of the total variance is explained (PC1 = 32.8%; PC2 = 19.2%) (Figure 1).

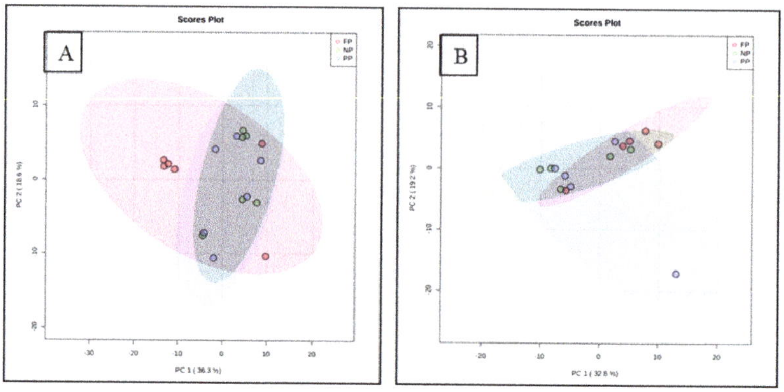

Figure 1. Principal component analysis (PCA) of meat (**A**) and subcutaneous fat; (**B**) metabolome of males programmed during gestation.

3.2. Heatmaps

Heatmaps allow you to visualize the global metabolomic profile of meat and fat and its variations between groups. And they represent a visual summary of the metabolites affected by treatments. The differentially expressed metabolites between the groups can be seen from Figure S1 (meat) and Figure S2 (fat) below.

3.3. Meat Metabolites

From the set of 188 metabolites evaluated, 25 metabolites were differentially expressed between treatments in meat ($p \leq 0.05$; 18 metabolites belonging to the class of phosphatidylcholines, 3 of amino acids, 2 of sphingolipids, 1 of biogenic amine, and 1 of lysophosphatidylcholine; (Table 1).

Table 1. Significant metabolites in meat from male cattle supplemented during pregnancy.

Metabolites	NP	PP	FP	p Value [1]
PC aa C26:0	0.65 ± 0.14^{a}	0.78 ± 0.13^{a}	0.36 ± 0.21^{b}	0.003
PC ae C34:2	3.17 ± 1.39^{ab}	5.91 ± 2.89^{a}	1.35 ± 1.71^{b}	0.010
PC ae C40:1	0.19 ± 0.13^{ab}	0.41 ± 0.18^{a}	0.14 ± 0.18^{b}	0.012
PC ae C34:1	6.33 ± 2.26^{a}	4.70 ± 2.01^{ab}	1.62 ± 2.06^{b}	0.013
PC aa C34:2	100.38 ± 60.14^{a}	94.48 ± 33.05^{a}	18.24 ± 26.06^{b}	0.017
PC ae C38:5	3.77 ± 1.87^{ab}	5.03 ± 2.51^{a}	1.33 ± 1.82^{b}	0.017
SM C18:0	10.25 ± 4.15^{ab}	14.15 ± 3.40^{a}	4.55 ± 5.69^{b}	0.019
Espermidine	0.35 ± 0.09^{b}	0.49 ± 0.13^{a}	0.52 ± 0.06^{a}	0.022
Ala	864.56 ± 118.87^{b}	1061.35 ± 74.07^{a}	981.71 ± 147.33^{ab}	0.022
SM (OH) C24:1	0.59 ± 0.17^{a}	0.41 ± 0.16^{ab}	0.22 ± 0.24^{b}	0.024
PC ae C36:5	3.77 ± 1.55^{ab}	5.35 ± 2.66^{a}	1.58 ± 2.09^{b}	0.025
PC aa C38:0	1.35 ± 0.55^{a}	1.28 ± 0.38^{a}	0.45 ± 0.66^{b}	0.026
PC ae C36:1	9.42 ± 3.51^{ab}	11.49 ± 6.88^{a}	2.82 ± 4.13^{b}	0.027
lisoPC a C26:1	0.29 ± 0.13^{a}	0.30 ± 0.10^{a}	0.10 ± 0.15^{b}	0.031
Thr	61.53 ± 9.96^{b}	80.00 ± 10.63^{a}	78.21 ± 11.78^{a}	0.032
PC ae C40:6	1.23 ± 0.56^{ab}	1.78 ± 0.79^{a}	0.60 ± 0.88^{b}	0.034
PC aa C32:2	7.42 ± 3.39^{ab}	10.42 ± 7.10^{a}	1.91 ± 3.52^{b}	0.035
PC aa C42:0	0.06 ± 0.03^{ab}	0.09 ± 0.04^{a}	0.04 ± 0.03^{b}	0.038
PC aa C34:3	8.94 ± 4.69^{a}	7.60 ± 3.76^{ab}	2.12 ± 2.97^{b}	0.040
PC ae C32:1	1.89 ± 0.95^{a}	1.60 ± 0.80^{ab}	0.48 ± 0.59^{b}	0.040
PC ae C36:2	19.28 ± 11.68^{a}	19.43 ± 5.84^{a}	5.16 ± 8.17^{b}	0.041
PC aa C36:2	288.65 ± 162.68^{a}	180.22 ± 118.81^{ab}	64.00 ± 94.01^{b}	0.043
Arg	45.40 ± 11.59^{b}	59.03 ± 8.93^{a}	59.62 ± 9.35^{a}	0.044
PC ae C44:6	0.11 ± 0.04^{ab}	0.16 ± 0.05^{a}	0.09 ± 0.04^{b}	0.045
PC ae C36:4	4.66 ± 3.02^{ab}	7.54 ± 4.99^{a}	1.29 ± 2.01^{b}	0.049

[1]—p value between treatments. The small letters superscripted represent the significant contrasts. NP—not programmed; PP—partially programmed; FP—full programming.

Regarding the class of phosphatidylcholines, the metabolites PC aa C26:0; PC aa C34:2; PC aa C38:0; and PC ae C36:2 had lower concentrations in the FP treatment compared to the NP and PP treatments ($p \leq 0.05$). Conversely, in the PP treatment, there was a higher concentration compared to the FP treatment for the following metabolites: PC ae C34:2; PC ae C40:1; PC ae C38:5; PC ae C36:5; PC ae C36:1; PC ae C40:6; PC aa C32:2; PC aa C42:0; PC ae C44:6; and PC ae C36:4 ($p \leq 0.05$). The metabolites PC ae C34:1; PC aa C34:3; PC ae C32:1; and PC aa C36:2 showed a lower concentration in the FP treatment compared to the NP treatment ($p \leq 0.05$).

In the amino acid class, the threonine ($p = 0.032$) and arginine ($p = 0.044$) levels were higher in treatments PP and FP, but alanine ($p = 0.022$) was higher in the PP group in relation to the NP group. In the class of sphingolipids, SM C18:0 ($p = 0.018$) was lower in FP compared to PP and SM (OH) C24:1 ($p = 0.024$) lower in FP compared to NP. For the metabolite lysoPC at C26:1 ($p = 0.031$), belonging to the lysophosphatidylcholine class, a smaller amount was found in FP, relative to NP and PP in meat. In the class of biogenic amines, spermidine ($p = 0.022$) had a higher expression in the treated groups (PP and FP).

3.4. Subcutaneous Fat Metabolites

From the set of 188 metabolites evaluated in this study, 10 metabolites were differentially expressed between treatments in the subcutaneous fat ($p \leq 0.05$). These different metabolites comprise nine in the phosphatidylcholine class and one in lysophosphatidylcholine (Table 2).

Table 2. Significant metabolites of subcutaneous fat from males submitted to fetal programming.

Metabolites	NP	PP	FP	p Value [1]
PC aa C36:0	2.06 ± 2.89^{ab}	1.48 ± 1.01^{b}	7.39 ± 5.11^{a}	0.013
PC ae C40:5	0.87 ± 0.84^{b}	1.19 ± 0.77^{ab}	2.32 ± 1.09^{a}	0.016
PC ae C36:0	1.64 ± 2.52^{ab}	0.77 ± 0.49^{b}	4.43 ± 2.88^{a}	0.019
PC ae C38:0	1.32 ± 1.51^{ab}	1.08 ± 0.64^{b}	3.24 ± 1.55^{a}	0.027
PC aa C38:3	7.23 ± 9.04^{b}	7.83 ± 8.98^{ab}	23.52 ± 14.35^{a}	0.031
PC aa C38:5	7.12 ± 8.26^{b}	9.12 ± 8.92^{ab}	19.84 ± 12.54^{a}	0.032
PC aa C36:1	43.79 ± 43.77^{b}	35.70 ± 33.50^{b}	143.48 ± 83.53^{a}	0.034
PC ae C30:2	0.40 ± 0.35^{b}	0.41 ± 0.35^{b}	1.11 ± 0.49^{a}	0.042
PC aa C40:6	2.65 ± 3.74^{b}	2.92 ± 3.77^{ab}	9.15 ± 8.32^{a}	0.048
lysoPC a C26:1	0.11 ± 0.10^{b}	0.25 ± 0.15^{a}	0.21 ± 0.07^{ab}	0.049

[1]—p value between treatments. The small letters superscripted represent the significant contrasts. NP—not programmed; PP—partially programmed; FP—full programming.

Regarding the metabolites PC ae C40:5; PC aa C38:3; PC aa C38:5; and PC aa C40:6 of the phosphatidylcholine class, the values obtained from the FP treatment were higher than from NP ($p \leq 0.05$). As for PC aa C36:0; PC ae C36:0; and PC ae C38:0, the FP treatment showed higher values than the PP treatment ($p \leq 0.05$). A significant difference for PC aa C36:1 and PC ae C30:2 was found in the higher expression with FP compared to with NP and PP ($p \leq 0.05$).

In the lysophosphatidylcholine class, lysoPC at C26:1 ($p = 0.029$) had its expression increased in PP compared to the control group (NP), but did not differ from FP.

3.5. Enrichment Analysis

The aminoacyl–transporter RNA biosynthesis pathway was significant ($p = 0.03$) only for the metabolites present in meat from the functional enrichment analysis of metabolic concentration data using the Kyoto Encyclopedia of Genes and Genomes (KEGG) pathway (Figure 2). For the other processes, it was not possible to find a difference between the groups; even though the ratio of enrichment of the valine, leucine, and isoleucine biosynthesis pathway proved to be greater than the aminoacyl–tRNA biosynthesis, it was not enough to present a significant p value.

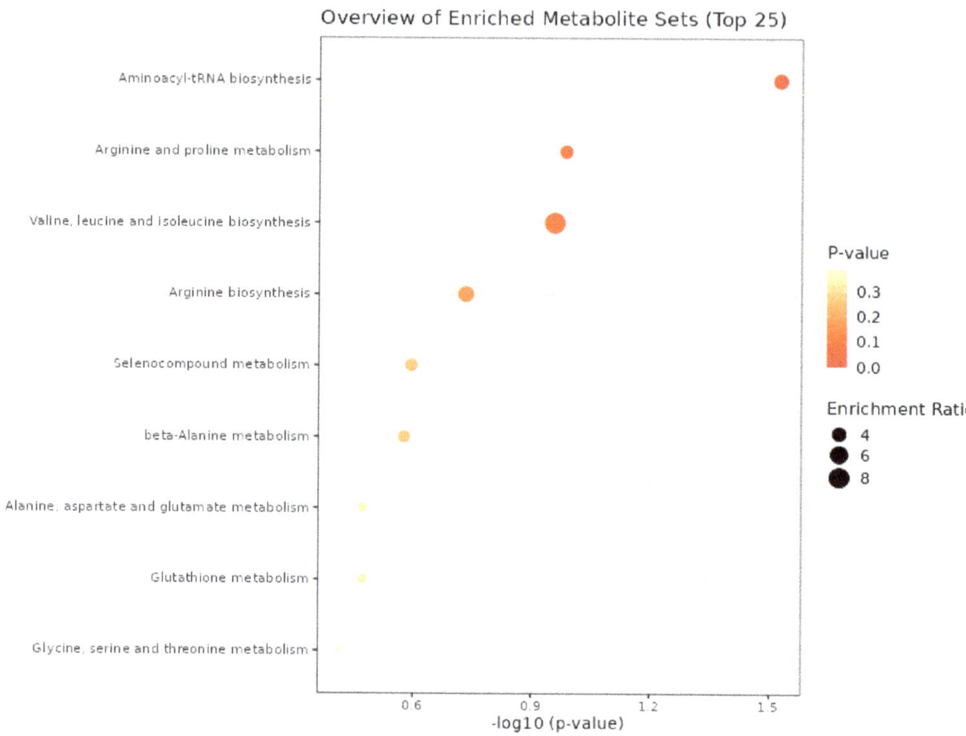

Figure 2. Metabolic processes expressed in the meat of males submitted to fetal programming (NP, PP and FP).

4. Discussion

There were many altered metabolites in the tissues and the classes to which they belonged were selected for discussion.

Amino acids are the small units that comprise proteins and protein is the most valuable meat component from a nutritional and marketing viewpoint [35]. Threonine, arginine, and leucine are examples of the amino acids in beef [36].

In our study, we found significance in the threonine and arginine levels, as we obtained a greater expression of these metabolites in treatments with prenatal nutritional stimulation, indicating that supplementation during pregnancy alters the profile of Thr and Arg in the meat of male progeny. As for the third amino acid found, alanine (Ala), we observed a difference only in the group that received supplementation in the final third of pregnancy (PP treatment).

Corroborating our results, Kwon et al. (2004) [37] found that food restriction throughout pregnancy causes a decrease in the same three amino acids mentioned above (Thr, Arg, and Ala). This shows that nutrient prenatal supplementation during gestation impacts the amino acid metabolism of the offspring at slaughter. Similarly, Hellmuth et al. (2016) [38] found effects of maternal nutritional restriction on the energy metabolism in baboons, with increased hepatic glucose concentration and decreased plasma levels of the amino acids methionine and threonine and hepatic threonine in the offspring that underwent restriction.

The class of sphingolipids is the second largest in terms of membrane lipids in animals and plants [39]. The metabolites SM C18:0 and SM(OH) C24:1 are part of a subcategory of sphingolipids called sphingomyelin. Sphingomyelins participate in cell proliferation, extracellular and intracellular signaling, cell differentiation, autophagy, and apoptosis [40]. From a nutritional viewpoint, the presence of sphingolipids in meat inhibits colon carcinogenesis, decreases serum LDL cholesterol, and raises HDL, suggesting that these metabolites represent a functional food constituent [41,42]. Our results show that prenatal supplementation throughout the entire gestation (FP treatment) increased this class of metabolites in relation to the control group or the PP treatment. Therefore, constant stimulus with nutrients during pregnancy ensures the benefits of this compound for the animal and for the consumer.

Phosphatidylcholines and lysophosphatidylcholines are part of the class of Glycerophospholipids, which is the main constituent of the mitochondrial membrane. This means that the more these metabolites are found in meat or fat, the more mitochondria are likely to be found [43]. Consequently, there may be a greater oxidation of these tissues since the function of mitochondria is to promote oxidation to obtain energy. The quality of a meat product with a large number of glycerophospholipids and that remains exposed on meat shelves at retail stores for a longer time may have its quality compromised by the oxidation process of lipids and by a rancid flavor [44]. In our study, the meat of the animals from the control group contained a larger number of glycerophospholipids. In the subcutaneous fat, however, we see the opposite, with a greater presence of glycerophospholipids in the fat of animals with prenatal nutritional stimulation.

Similarly, Muroya et al. (2021) [29] found changes in the glycerophospholipid expression pathway in the muscle of bovine fetuses from malnutrition during pregnancy. These modifications may interfere with postnatal muscle growth and repair [45].

Biogenic amines, such as spermidine, are metabolites formed by nitrogenous bases from the decarboxylation of free amino acids by microbial action, and are considered antinutritional [46]. This classification has two important motivations in terms of meat consumers since the food can have different levels of biogenic amines [39]. First, the health risk to humans due to the toxicity of biogenic amines and possible drug interactions is an issue [47,48]. Second, the quality and acceptability of the food are compromised [49]. As the level of amines depends on the action of bacteria, which can be toxic, the metabolite can be observed to indicate food freshness. The greater the production of biogenic amines, the larger the number of microorganisms and the more deteriorated the food becomes [50]. Furthermore, spermidine, which is categorized as a polyamine, can be altered by physiological stimuli or by inhibitors of metabolic enzymes [51]. Our work indicated that prenatal nutritional stimulus of cows increased the meat spermidine concentration of male offspring, which may indicate a possible worsening of meat quality. In pregnant cows that underwent nutritional restriction, an increase in the spermidine precursor was also observed, possibly altering the metabolism of this polyamine in the offspring [29].

Functional enrichment analysis aims to identify changes in the expression of pathways that comprise the biological processes, both when increased and decreased or when associated with the physiological responses of organisms. In the present study, nine processes were found; however, only one was significant among the treatments, with increased expression in the aminoacyl–transporter RNA biosynthesis pathway.

Biosynthesis is a chemical process that involves the production of various molecules in a living organism, such as lipids, nucleic acids, and proteins, from simple compounds. Simple reactants cluster to convert another compound or to build macromolecules [52]. The union/conversion process of these molecules comes from metabolic pathways. Metabolic pathways are responsible for receiving and transmitting substrates to other reactions; therefore, they are dependent on other pathways to occur. The aminoacyl–tRNA molecule is a transport enzyme that, along with the translation factors, carries amino acids to the ribosomes that incorporate these amino acids into peptide chains for protein formation [53]. Thus, the increase in the pathway responsible for the biosynthesis of aminoacyl–tRNA translates into more reactions occurring in the meat of the animals in the PP and FP treatments for the formation of aminoacyl–tRNA enzymes, which is positive since the latter have the function of providing varied proteins to the cell membrane, such as the proteins that make up muscle. The process becomes even more advantageous when the activation of metabolic pathways, such as this one, happens during pregnancy, as the moment of formation and growth of new tissues happens in that period and can modulate the results throughout the animal's life [54].

The discoveries conferred by the use of new technologies in the present study offer a better understanding of physiological processes and provide excellent advances in beef production as a whole [14]. The continuous development of new technologies and methods for complex data analyses using modeling and prediction of functional biochemical networks allows us to approach the general biology that combines molecular, genetic, and environmental data, contributing to better understanding the phenotype of interest [55].

5. Conclusions

The prenatal nutrition of Nellore cows was able to modify the levels of some metabolites present in the meat and subcutaneous adipose tissue of the male offspring. Furthermore, a significant metabolic process was found in the meat metabolome to be correlated with different supplementation strategies during pregnancy (the aminoacyl–transporter RNA biosynthesis pathway). Thus, maternal nutrition can modulate pathways and metabolites that culminate in phenotypic changes in meat and fat, which can alter the animal's performance characteristics and/or the quality of the final product for the consumer.

Supplementary Materials: The following supporting information can be downloaded at https://www.mdpi.com/article/10.3390/metabo14010009/s1, Figure S1: Metabolomic profile of meat from bulls supplemented during pregnancy; Figure S2: Metabolomic profile of subcutaneous fat in bulls that received prenatal nutrition.

Author Contributions: Conceptualization, J.B.S.F.; formal analysis, G.H.G.P.; funding acquisition, M.H.d.A.S.; investigation, R.C.C.; methodology, G.C.B.; project administration, M.H.d.A.S.; software, F.A.C.Q.C.; supervision, M.D.P.; writing—review and editing, A.C.F. and M.H.d.A.S. All authors have read and agreed to the published version of the manuscript.

Funding: This research was funded by the São Paulo Research Foundation (FAPESP) [grant numbers 21/03802-7, 17/12105–2] and the National Council for Scientific and Technological Development (CNPq) [grant number 307593/2021-5].

Institutional Review Board Statement: This study was approved by the Research Ethics Committee of FZEA-USP under protocol no. 1843241117, on 10 March 2018, according to the guidelines of the National Council for the Control of Animal Experimentation (CONCEA).

Informed Consent Statement: Not applicable.

Data Availability Statement: None of the data were deposited into an official repository. The raw data supporting the conclusion of this article will be made available by the authors without undue reservation.

Acknowledgments: The authors acknowledge the College of Animal Science and Food Engineering (FZEA-USP) for providing all the human and infrastructure resources that made this study possible.

Conflicts of Interest: The authors declare that the research was conducted in the absence of any commercial or financial relationships that could be construed as potential conflicts of interest. The reviewer J.P. declares a shared affiliation with the authors to the handling editor at the time of the review. The authors declare no conflict of interest.

References

1. Nowacka-Woszuk, J. Nutrigenomics in Livestock—Recent Advances. *J. Appl. Genet.* **2020**, *61*, 93–103. [CrossRef]
2. Kauffman, R.G. Meat Composition. In *Handbook of Meat and Meat Processing*, 2nd ed.; Routledge: London, UK, 2012; pp. 45–61. [CrossRef]
3. Costa, T.C.; Gionbelli, M.P.; de Souza Duarte, M. Fetal Programming in Ruminant Animals: Understanding the Skeletal Muscle Development to Improve Meat Quality. *Anim. Front.* **2021**, *11*, 66–73. [CrossRef]
4. Long, N.M.; Prado-Cooper, M.J.; Krehbiel, C.R.; Desilva, U.; Wettemann, R.P. Effects of Nutrient Restriction of Bovine Dams during Early Gestation on Postnatal Growth, Carcass and Organ Characteristics, and Gene Expression in Adipose Tissue and Muscle. *J. Anim. Sci.* **2010**, *88*, 3251–3261. [CrossRef]

5. Emwas, A.H.M.; Al-Rifai, N.; Szczepski, K.; Alsuhaymi, S.; Rayyan, S.; Almahasheer, H.; Jaremko, M.; Brennan, L.; Lachowicz, J.I. You Are What You Eat: Application of Metabolomics Approaches to Advance Nutrition Research. *Foods* **2021**, *10*, 1249. [CrossRef]
6. Werner, E.; Heilier, J.F.; Ducruix, C.; Ezan, E.; Junot, C.; Tabet, J.C. Mass Spectrometry for the Identification of the Discriminating Signals from Metabolomics: Current Status and Future Trends. *J. Chromatogr. B* **2008**, *871*, 143–163. [CrossRef]
7. Griffin, J.L. The Cinderella Story of Metabolic Profiling: Does Metabolomics Get to Go to the Functional Genomics Ball? *Philos. Trans. R. Soc. B Biol. Sci.* **2005**, *361*, 147–161. [CrossRef]
8. McCoard, S.A.; Stevens, D.R.; Whitney, T.R. Sustainable Sheep and Goat Production through Strategic Nutritional Management and Advanced Technologies. *Anim. Agric. Sustain. Chall. Innov.* **2020**, 231–246. [CrossRef]
9. Goldansaz, S.A.; Guo, A.C.; Sajed, T.; Steele, M.A.; Plastow, G.S.; Wishart, D.S. Livestock Metabolomics and the Livestock Metabolome: A Systematic Review. *PLoS ONE* **2017**, *12*, e0177675. [CrossRef] [PubMed]
10. Picard, B.; Lebret, B.; Cassar-Malek, I.; Liaubet, L.; Berri, C.; Le Bihan-Duval, E.; Hocquette, J.F.; Renand, G. Recent Advances in Omic Technologies for Meat Quality Management. *Meat Sci.* **2015**, *109*, 18–26. [CrossRef] [PubMed]
11. Berri, C.; Picard, B.; Lebret, B.; Andueza, D.; Lefèvre, F.; Le Bihan-Duval, E.; Beauclercq, S.; Chartrin, P.; Vautier, A.; Legrand, I.; et al. Predicting the Quality of Meat: Myth or Reality? *Foods* **2019**, *8*, 436. [CrossRef] [PubMed]
12. Carrillo, J.A.; He, Y.; Li, Y.; Liu, J.; Erdman, R.A.; Sonstegard, T.S.; Song, J. Integrated Metabolomic and Transcriptome Analyses Reveal Finishing Forage Affects Metabolic Pathways Related to Beef Quality and Animal Welfare. *Sci. Rep.* **2016**, 6. [CrossRef] [PubMed]
13. Yu, K.; Matzapetakis, M.; Valent, D.; Saco, Y.; De Almeida, A.M.; Terré, M.; Bassols, A. Skeletal Muscle Metabolomics and Blood Biochemistry Analysis Reveal Metabolic Changes Associated with Dietary Amino Acid Supplementation in Dairy Calves. *Sci. Rep.* **2018**, *8*. [CrossRef] [PubMed]
14. Zhang, T.; Chen, C.; Xie, K.; Wang, J.; Pan, Z. Current State of Metabolomics Research in Meat Quality Analysis and Authentication. *Foods* **2021**, *10*, 2388. [CrossRef] [PubMed]
15. Antonelo, D.; Gómez, J.F.M.; Cônsolo, N.R.B.; Beline, M.; Colnago, L.A.; Schilling, W.; Zhang, X.; Suman, S.P.; Gerrard, D.E.; Balieiro, J.C.C.; et al. Metabolites and Metabolic Pathways Correlated With Beef Tenderness. *Meat Muscle Biol.* **2020**, *4*, 19–20. [CrossRef]
16. Zuo, H.; Wang, P.; Guo, Z.; Luo, X.; Zhang, Y.; Mao, Y. Metabolites Analysis on Water-Holding Capacity in Beef Longissimus Lumborum Muscle during Postmortem Aging. *Metabolites* **2022**, *12*, 242. [CrossRef]
17. Bischof, G.; Witte, F.; Terjung, N.; Heinz, V.; Juadjur, A.; Gibis, M. Metabolic, Proteomic and Microbial Changes Postmortem and during Beef Aging. *Crit. Rev. Food Sci. Nutr.* **2022**, 1–34. [CrossRef]
18. Muroya, S.; Ueda, S.; Komatsu, T.; Miyakawa, T.; Ertbjerg, P. MEATabolomics: Muscle and Meat Metabolomics in Domestic Animals. *Metabolites* **2020**, *10*, 188. [CrossRef]
19. Du, M.; Tong, J.; Zhao, J.; Underwood, K.R.; Zhu, M.; Ford, S.P.; Nathanielsz, P.W. Fetal Programming of Skeletal Muscle Development in Ruminant Animals. *J. Anim. Sci.* **2010**, *88*, E51–E60. [CrossRef]
20. Du, M.; Zhao, J.X.; Yan, X.; Huang, Y.; Nicodemus, L.V.; Yue, W.; Mccormick, R.J.; Zhu, M.J. Fetal Muscle Development, Mesenchymal Multipotent Cell Differentiation, and Associated Signaling Pathways. *J. Anim. Sci.* **2011**, *89*, 583–590. [CrossRef]
21. de Zawadzki, A.; Arrivetti, L.O.R.; Vidal, M.P.; Catai, J.R.; Nassu, R.T.; Tullio, R.R.; Berndt, A.; Oliveira, C.R.; Ferreira, A.G.; Neves-Junior, L.F.; et al. Mate Extract as Feed Additive for Improvement of Beef Quality. *Food Res. Int.* **2017**, *99*, 336–347. [CrossRef]
22. Peng, D.Q.; Kim, S.J.; Lee, H.G. Metabolomics Analyses to Characterize Metabolic Alterations in Korean Native Calves by Oral Vitamin A Supplementation. *Sci. Rep.* **2020**, *10*, 8092. [CrossRef] [PubMed]
23. Gómez, J.F.M.; Cônsolo, N.R.B.; Antonelo, D.S.; Beline, M.; Gagaoua, M.; Higuera-Padilla, A.; Colnago, L.A.; Gerrard, D.E.; Silva, S.L. Impact of Cattle Feeding Strategy on the Beef Metabolome. *Metabolites* **2022**, *12*, 640. [CrossRef] [PubMed]
24. Cônsolo, N.R.B.; Buarque, V.L.M.; Silva, J.; Poleti, M.D.; Barbosa, L.C.G.S.; Higuera-Padilla, A.; Gómez, J.F.M.; Colnago, L.A.; Gerrard, D.E.; Saran Netto, A.; et al. Muscle and Liver Metabolomic Signatures Associated with Residual Feed Intake in Nellore Cattle. *Anim. Feed Sci. Technol.* **2021**, *271*, 114757. [CrossRef]
25. Gandra, J.R.; Pedrini, C.A.; Cônsolo, N.R.B.; Acosta, A.P.; Seno, L.O.; Barbosa, L.C.G.S.; Noia, I.Z.; Buarque, V.L.M.; Padilla, A.R.H.; Colnago, L.A.; et al. Metabolome Fingerprints, Performance and Carcass Quality of Beef Calves Supplemented with Antibiotic Free Additive. *Anim. Biotechnol.* **2020**, *33*, 710–722. [CrossRef]
26. Schalch Junior, F.J.; Polizel, G.H.G.; Cançado, F.A.C.Q.; Fernandes, A.C.; Mortari, I.; Pires, P.R.L.; Fukumasu, H.; Santana, M.H.d.A.; Saran Netto, A. Prenatal Supplementation in Beef Cattle and Its Effects on Plasma Metabolome of Dams and Calves. *Metabolites* **2022**, *12*, 347. [CrossRef] [PubMed]
27. Polizel, G.H.G.; Cançado, F.A.C.Q.; Dias, E.F.F.; Fernandes, A.C.; Cracco, R.C.; Carmona, B.T.; Castellar, H.H.; Poleti, M.D.; de Almeida Santana, M.H. Effects of Different Prenatal Nutrition Strategies on the Liver Metabolome of Bulls and Its Correlation with Body and Liver Weight. *Metabolites* **2022**, *12*, 441. [CrossRef]
28. Martin, D.E.; Jones, A.K.; Pillai, S.M.; Hoffman, M.L.; McFadden, K.K.; Zinn, S.A.; Govoni, K.E.; Reed, S.A. Maternal Restricted- And Overfeeding during Gestation Result in Distinct Lipid and Amino Acid Metabolite Profiles in the Longissimus Muscle of the Offspring. *Front. Physiol.* **2019**, *10*, 515. [CrossRef]

29. Muroya, S.; Zhang, Y.; Kinoshita, A.; Otomaru, K.; Oshima, K.; Gotoh, Y.; Oshima, I.; Sano, M.; Roh, S.; Oe, M.; et al. Maternal Undernutrition during Pregnancy Alters Amino Acid Metabolism and Gene Expression Associated with Energy Metabolism and Angiogenesis in Fetal Calf Muscle. *Metabolites* **2021**, *11*, 582. [CrossRef]
30. National Research Council. *Nutrient Requirements of Beef Cattle: Seventh Revised Edition: Update 2000*, 7th ed.; The National Academies Press: Washington, DC, USA, 2000.
31. Ministério da Agricultura e Pecuária—Governo Federal Brasileiro DECRETO No 30.691, de 29 de MARÇO de 1952. Regulamento da Inspeção Industrial e Sanitária de Produtos de Origem Animal. 1952, pp. 1–44. Available online: https://www2.camara.leg.br/legin/fed/decret/1950-1959/decreto-30691-29-marco-1952-339586-normaatualizada-pe.pdf (accessed on 10 December 2023).
32. Ludtke, C.B.; Ciocca, J.R.P.; Dandin, T.; Barbalho, P.C.; Vilela, J.A.; Ferrarini, C.; Abate Humanitário de Bovinos. AMA Editora e Treinamento Ltda, Rio de Janeiro. 2012; p. 148. Available online: https://www.gov.br/agricultura/pt-br/assuntos/producao-animal/arquivos-publicacoes-bem-estar-animal/programa-steps-2013-abate-humanitario-de-bovinos.pdf (accessed on 10 December 2023).
33. Cracco, R.C.; Ruy, I.M.; Polizel, G.H.G.; Fernandes, A.C.; Furlan, É.; Baldin, G.C.; Santos, G.E.C.; Santana, M.H. de A. Evaluation of Maternal Nutrition Effects in the Lifelong Performance of Male Beef Cattle Offspring. *Vet. Sci.* **2023**, *10*, 443. [CrossRef]
34. Zukunft, S.; Prehn, C.; Röhring, C.; Möller, G.; Hrabě de Angelis, M.; Adamski, J.; Tokarz, J. High-Throughput Extraction and Quantification Method for Targeted Metabolomics in Murine Tissues. *Metabolomics* **2018**, *14*, 18. [CrossRef]
35. Geletu, U.S.; Usmael, M.A.; Mummed, Y.Y.; Ibrahim, A.M. Quality of Cattle Meat and Its Compositional Constituents. *Vet. Med. Int.* **2021**, *2021*, 7340495. [CrossRef] [PubMed]
36. Olaoye, O.A. MiniReview Meat: An Overview of Its Composition, Biochemical Changes and Associated Microbial Agents. *Int. Food Res. J.* **2011**, *18*, 877–885.
37. Kwon, H.; Ford, S.P.; Bazer, F.W.; Spencer, T.E.; Nathanielsz, P.W.; Nijland, M.J.; Hess, B.W.; Wu, G. Maternal Nutrient Restriction Reduces Concentrations of Amino Acids and Polyamines in Ovine Maternal and Fetal Plasma and Fetal Fluids. *Biol. Reprod.* **2004**, *71*, 901–908. [CrossRef] [PubMed]
38. Hellmuth, C.; Uhl, O.; Kirchberg, F.F.; Harder, U.; Peissner, W.; Koletzko, B.; Nathanielsz, P.W. Influence of Moderate Maternal Nutrition Restriction on the Fetal Baboon Metabolome at 0.5 and 0.9 Gestation. *Nutr. Metab. Cardiovasc. Dis.* **2016**, *26*, 786–796. [CrossRef] [PubMed]
39. Ruiz-Capillas, C.; Jiménez-Colmenero, F. Biogenic Amines in Meat and Meat Products. *Crit. Rev. Food Sci. Nutr.* **2010**, *44*, 489–599. [CrossRef] [PubMed]
40. Hannun, Y.A.; Obeid, L.M. Principles of Bioactive Lipid Signalling: Lessons from Sphingolipids. *Nat. Rev. Mol. Cell Biol.* **2008**, *9*, 139–150. [CrossRef]
41. Vesper, H.; Schmelz, E.M.; Nikolova-Karakashian, M.N.; Dillehay, D.L.; Lynch, D.V.; Merrill, A.H. Sphingolipids in Food and the Emerging Importance of Sphingolipids to Nutrition. *J. Nutr.* **1999**, *129*, 1239–1250. [CrossRef]
42. Fischbeck, A.; Krüger, M.; Blaas, N.; Humpf, H.U. Analysis of Sphingomyelin in Meat Based on Hydrophilic Interaction Liquid Chromatography Coupled to Electrospray Lonization-Tandem Mass Spectrometry (HILIC-HPLC-ESI-MS/MS). *J. Agric. Food Chem.* **2009**, *57*, 9469–9474. [CrossRef]
43. Muroya, S. An Insight into Farm Animal Skeletal Muscle Metabolism Based on a Metabolomics Approach. *Meat Sci.* **2023**, *195*, 108995. [CrossRef]
44. Yu, Q.; Tian, X.; Shao, L.; Li, X.; Dai, R. Mitochondria Changes and Metabolome Differences of Bovine Longissimus Lumborum and Psoas Major during 24 h Postmortem. *Meat Sci.* **2020**, *166*, 108112. [CrossRef]
45. Nagata, Y.; Partridge, T.A.; Matsuda, R.; Zammit, P.S. Entry of Muscle Satellite Cells into the Cell Cycle Requires Sphingolipid Signaling. *J. Cell Biol.* **2006**, *174*, 245–253. [CrossRef]
46. Jairath, G.; Singh, P.K.; Dabur, R.S.; Rani, M.; Chaudhari, M. Biogenic Amines in Meat and Meat Products and Its Public Health Significance: A Review. *J. Food Sci. Technol.* **2015**, *52*, 6835–6846. [CrossRef]
47. Shalaby, A.R. Significance of Biogenic Amines to Food Safety and Human Health. *Food Res. Int.* **1996**, *29*, 675–690. [CrossRef]
48. Bardócz, S. Polyamines in Food and Their Consequences for Food Quality and Human Health. *Trends Food Sci. Technol.* **1995**, *6*, 341–346. [CrossRef]
49. Halász, A.; Baráth, Á.; Simon-Sarkadi, L.; Holzapfel, W. Biogenic Amines and Their Production by Microorganisms in Food. *Trends Food Sci. Technol.* **1994**, *5*, 42–49. [CrossRef]
50. Stadnik, J.; Dolatowski, Z.J. Corresponding Author-Adres Do Korespondencji: Dr Inż Biogenic Amines in Meat and Fermented Meat Products. *Acta Sci. Pol. Technol. Aliment* **2010**, *9*, 251–263.
51. Cervelli, M.; Leonetti, A.; Duranti, G.; Sabatini, S.; Ceci, R.; Mariottini, P. Skeletal Muscle Pathophysiology: The Emerging Role of Spermine Oxidase and Spermidine. *Med. Sci.* **2018**, *6*, 14. [CrossRef]
52. Ijaz, M.; Zhang, D.; Hou, C.; Mahmood, M.; Hussain, Z.; Zheng, X.; Li, X. Changes in Postmortem Metabolites Profile of Atypical and Typical DFD Beef. *Meat Sci.* **2022**, *193*, 108922. [CrossRef]
53. Zou, Y.; Yang, Y.; Fu, X.; He, X.; Liu, M.; Zong, T.; Li, X.; Htet Aung, L.; Wang, Z.; Yu, T. The Regulatory Roles of Aminoacyl-TRNA Synthetase in Cardiovascular Disease. *Mol. Ther. Nucleic Acids* **2021**, *25*, 372–387. [CrossRef]

54. Zhu, M.J.; Ford, S.P.; Means, W.J.; Hess, B.W.; Nathanielsz, P.W.; Du, M. Maternal Nutrient Restriction Affects Properties of Skeletal Muscle in Offspring. *J. Physiol.* **2006**, *575*, 241–250. [CrossRef]
55. Suravajhala, P.; Kogelman, L.J.A.; Kadarmideen, H.N. Multi-Omic Data Integration and Analysis Using Systems Genomics Approaches: Methods and Applications in Animal Production, Health and Welfare. *Genet. Sel. Evol.* **2016**, *48*. [CrossRef] [PubMed]

metabolites

MDPI

Article

Changes in the Lipid Metabolism of the *Longissimus thoracis* Muscle in Bulls When Using Different Feeding Strategies during the Growing and Finishing Phases

Juliana Akamine Torrecilhas [1], Guilherme Luis Pereira [1,*], Elias San Vito [2], Giovani Fiorentini [3], Germán Darío Ramirez-Zamudio [4], Larissa Simielli Fonseca [5], Rodrigo de Nazaré Santos Torres [1], Tiago Adriano Simioni [5], Juliana Messana Duarte [5], Otavio Rodrigues Machado Neto [1], Rogério Abdallah Curi [1], Luis Artur Loyola Chardulo [1], Welder Angelo Baldassini [1] and Telma Teresinha Berchielli [5]

1 School of Veterinary e Animal Science (FMVZ), São Paulo State University (Unesp), Jaboticabal 14884-900, SP, Brazil; juliana.akamine@unesp.br (J.A.T.); rodrigo.torres@unesp.br (R.d.N.S.T.); otavio.machado@unesp.br (O.R.M.N.); rogerio.curi@unesp.br (R.A.C.); luis.artur@unesp.br (L.A.L.C.); w.baldassini@unesp.br (W.A.B.)

2 Confina Beef Cattle Consulting, Sinop 78555-603, MT, Brazil; sanvitoelias@gmail.com

3 Department of Animal Science, Federal University of Pelotas (UFPEL), Pelotas 96160-000, RS, Brazil; giovani.fiorentini@ufpel.edu.br

4 College of Animal Science and Food Engineering (FZEA), University of São Paulo (USP), Pirassununga 13635-900, SP, Brazil; germanramvz@usp.br

5 School of Agriculture and Veterinary Sciences (FCAV), São Paulo State University (Unesp), Jaboticabal 14884-900, SP, Brazil; larissa.simielli@unesp.br (L.S.F.); simioni@zootecnista.com.br (T.A.S.); juliana.d.messana@unesp.br (J.M.D.); telma.berchielli@unesp.br (T.T.B.)

* Correspondence: guilherme.luis@unesp.br; Tel.: +55-(14)-98144-0486

Abstract: The objective was to evaluate the supplementation strategy's effect on beef cattle during the growing phase and two systems during the finishing phase. One hundred and twenty young bulls were randomly divided in a 2 × 2 factorial design to receive either mineral (ad libitum) or protein + energy (3 g/kg body weight (BW)/day) during the growing phase and pasture plus concentrate supplementation (20 g/kg BW/day) or feedlot (25:75% corn silage:concentrate) during the finishing phase. Feedlot-fed bulls had meat (*Longissimus thoracis*—LT) with a higher content of lipids and saturated and monounsaturated fatty acids and a greater upregulation of *stearoyl-CoA desaturase* and *sterol regulatory element-binding protein-1c* than animals that fed on pasture ($p < 0.05$). On the other hand, pasture-fed bulls had meat with a higher content of α-linoleic acid, linolenic acid, and n6 and a greater n6:n3 ratio compared to the feedlot-fed group ($p < 0.05$). In addition, meat from pasture-fed bulls during the finishing phase had 17.6% more isocitrate dehydrogenase enzyme concentration than the feedlot group ($p = 0.02$). Mineral-fed and pasture-finished bulls showed down-regulation of *peroxisome proliferator-activated receptor gamma* ($p < 0.05$), while the bulls fed protein + energy and finished in the feedlot had higher *carnitine palmitoyltransferase 2* expression ($p \leq 0.013$). In conclusion, mineral or protein + energy supplementation in the growing does not affect the fatty acid composition of intramuscular fat of LT muscle. In the finishing phase, feeding bulls in the feedlot upregulates the lipogenic genes and consequently improves the intramuscular fat content in the meat.

Keywords: beef cattle; lipogenic genes; marbling; meat quality

1. Introduction

In the growing phase of beef cattle, supplementation strategies are used to increase the efficiency of the grazing system and animal performance since tropical forages barely meet the nutritional requirements of animals. Furthermore, the growing phase is a critical point in reducing the slaughter age in response to better biological efficiency in tissue deposition in young animals [1]. In this sense, most studies have evaluated the effect of supplementation during the growing phase on the finishing performance of cattle [2–4]. However, little is known about how skeletal muscle metabolism is affected by these nutritional strategies and their response to meat quality, especially intramuscular fat deposition, a criterion of great importance in several countries.

Fat deposition is a response to the activation of metabolic pathways that control uptake, synthesis, and lipolysis, which occur according to the need for lipid release or storage and are regulated by the interaction of dietary nutrients and the expression level of genes involved in lipid metabolism [5]. Animal supplementation with concentrated diets may increase the amount of insulin in the blood, which stimulates glucose uptake by tissues, consequently increasing the amount of intramuscular lipids [6]. Such conditions could increase the expression of genes such as *SCD-1*, which is associated with the conversion of saturated fatty acids into monounsaturated. Greater expression of *SCD-1* is dependent on metabolic signals such as glucose and insulin in the blood, as reported [7].

The use of a high-concentrate diet in cattle feed is important during the finishing phase and can reduce the feedlot period and improve carcass fat deposition. Although the cattle finished in the pasture system supplemented with higher concentrate (1.5 to 2% of body weight [BW]) had higher nutrient requirements than those in the feedlot system [8], this system can be used alternatively to produce carcasses with minimum cover fat [9] as the operational costs of the system may decrease. However, meat from cattle supplemented with grain feed is known to have a greater amount of saturated fatty acids (SFAs) and a less favorable n6/n3 ratio than those from cattle fed exclusively with grass [10]. Although previous works have investigated the beef quality in different systems [11], there are no studies that evaluate the effect of supplementing with a high-concentrate diet (2% BW) on the intramuscular fat and fatty acid profile of meat from cattle finished in pasture systems (2% BW).

It was hypothesized that the supplementation strategy in the growing, followed by feedlot finishing, would influence skeletal muscle metabolism by regulating lipogenic genes, which may impact intramuscular fat deposition. In contrast, finishing bulls on pasture with high concentrate supplementation will have higher unsaturated fatty acid content and a more favorable n6/n3 ratio than animals finished in a feedlot system. In this context, the objective of this study was to evaluate the effect of the supplementation strategy during the growing and finishing phases (pasture supplemented with concentrate or conventional feedlot) on the fatty acid profile, lipogenic enzyme activity, and relative abundance of mRNA associated with lipid metabolism in the *Longissimus thoracis* (LT) muscle of beef cattle.

2. Materials and Methods

This study was approved by the Ethics and Animal Welfare Committee of São Paulo State University (protocol 5628/15). The study was carried out at the beef cattle facility of São Paulo State University, Jaboticabal, São Paulo, Brazil (48°1858′ W, 21°15 22′ S).

2.1. Animals

The experimental period comprised the growing phase (first experimental period) and the finishing phase (second experimental period). The study was conducted between December and September (285 d). One hundred and twenty bulls from three genetic groups: 40 Nellore (10 ± 2 months old; 264.80 ± 13.75 kg), 40 ½ Angus × ½ Nellore (11 ± 2 months old; 278.00 ± 20.32 kg), and 40 ½ Senepol × ½ Nellore (9 ± 2 months old; 226.70 ± 22.24 kg), were used. The animals were acquired from different herds, and due to this heterogeneity, we chose to use breed as a random effect and thus dilute the variation in response to the high sample number. Before the experiment, all bulls were fed grass without creep-feeding supplementation.

2.2. *Growing Phase (Growing Feed)*

The experiment was conducted in a randomised block design with two supplements, (1) mineral (ad libitum; $n = 60$) and (2) protein + energy (3 g/kg BW/day; $n = 60$), during the growing. The growing phase occurred during the summer season in Brazil (December to May, 155 d). At the beginning of the experiment, the bulls were divided based on BW and placed in one of two treatments during the growing. The supplements were based on tropical conditions [12], and the composition of the diets is presented in Table 1. The mineral premix was added in both treatments (mineral and protein + energy). The amount of supplement was calculated to meet the requirements for an average daily gain of 0.6 kg/d, according to Valadares Filho et al. [13]. During the growing phase, bulls were fed once a day (10:00 h), and the grazing area consisted of Brachiaria grass (*Urochloa brizantha cv.* "Xaraés") distributed into 12 paddocks (approximately 1.8 ha each), with 10 bulls/paddock. Each paddock had semi-circular drinkers and covered feed troughs (3.0 m × 0.5 m), with easy access to both sides for supplementation. Every 28 days, the bulls were weighed, and their BW was used to adjust the amount of supplement supplied. Mineral feed was available to the bulls ad libitum, and the protein + energy supplement amount was calculated based on BW at the beginning of each experimental period.

Table 1. Chemical composition of the experimental diets.

Items	Growing		Finishing		
	PRE [4]	Pasture	Concentrate	Pasture	Corn Silage
Ingredients, g/kg DM					
Corn	735	-	7890	-	-
Soybean meal	106	-	1650	-	-
Mineral premix [1]	159	-	46.00	-	-
Chemical composition					
Dry matter	860	332	899	458	301
Organic matter	892	925	910	923	950
Crude protein	205	128	160	113	95.0
Neutral detergent fibre	265	575	251	582	331
Ether extract	63.0	24.1	66.0	23.0	71.0
Fatty acid, g/100 g of total FA [3]					
Myristic (C14:0)	0.08	1.30	0.08	3.09	0.27
Palmitic (C16:0)	11.2	36.5	11.2	35.3	17.6
Margaric (C17:0)	0.09	0.49	0.09	0.65	0.22
Stearic (C18:0)	3.94	3.60	3.76	4.16	3.46
C20:0	0.38	0.99	0.38	1.71	0.84
C22:0	0.45	1.28	0.44	2.03	0.44
C24:0	0.18	2.12	0.20	3.02	0.78
Palmitoleic (C16:1)	0.12	0.46	0.09	0.42	0.23
Oleic (C18:1n9c)	28.4	4.35	29.9	6.17	34.2
Linoleic (C18:2n6c)	48.6	14.8	47.6	14.1	36.5
α-Linolenic (C18:3n3)	4.62	30.3	4.40	22.6	3.69
SFA [2]	16.3	46.3	16.1	49.9	23.7
MUFA [2]	28.5	4.81	29.9	6.59	34.4
PUFA [2]	53.2	45.1	52.1	36.6	40.2

[1] Sodium 80 g/kg; Calcium 153 g/kg; Phosphorus 30 g/kg; Sulfur 30 g/kg; Zinc 1925 mg/kg; Copper 520 mg/kg; Manganese 400 mg/kg; Iodine 30 mg/kg; Cobalt 38 mg/kg; Selenium 10 mg/kg; Vit A 55,000 Ul/kg; Vit D3 7500 Ul/kg; Vit E 750 Ul/kg; Monensin 400 mg/kg; NNP 620 g /kg; [2] SFA = saturated fatty acid; MUFA = monounsaturated fatty acids; PUFA = polyunsaturated fatty acids. [3] Fatty acid; [4] Protein + energy (3 g/kg BW/day).

2.3. *Finishing Phase (Second Experimental Phase)*

The finishing phase was conducted during the winter and dry seasons (May to September; 129 days). The adaptation period was 20 days, based on the "step-up" procedure, before the second experimental phase. Following the growing (first experimental phase),

30 bulls within each treatment (supplementation) were selected and assigned to one of the two finishing systems: (1) pasture plus concentrate supplementation (20 g/kg BW/day) and (2) feedlot system, where bulls received corn silage as roughage and concentrate (25:75; corn silage:concentrate). The chemical composition and profile of the fatty acids in the experimental diets are shown in Table 1. The amount of supplement was calculated to meet the requirements for an average daily gain of 1.5 kg/d, according to Valadares Filho et al. [13].

All bulls designated for treatment with pasture + supplementation were housed in the same paddock during the finishing phase (12 paddocks, with 5 bulls/paddock from growing feed). Every 28 days, the bulls were weighed, and their BW was used to adjust the amount of concentrate supplemented (20 g/kg BW/day). The bulls of the pasture system were fed concentrate once per day (10:00 h) during the experimental period. The bulls designated for the feedlot system were retained in individual pens of 12 m^2, partially covered concrete floors, with feed-trough and free water access. The feedlot basal diet comprised 750 g/kg concentrate (corn, soybean meal, and premix) and 250 g/kg roughage (corn silage). The bulls were fed twice per day (08:00 h and 15:00 h), and the amount of diet was adjusted weekly for a 5% feed refusals.

2.4. Slaughter Procedure and Muscle Sampling

After 285 days, bulls with BW of 510.90 ± 43.65 kg (Nellore), 532.70 ± 55.81 kg (½ Angus) and 466.20 ± 48.48 kg (½ Senepol) were transported to a commercial slaughterhouse (Minerva Foods, Barretos, São Paulo, Brazil) located 90 km from the experimental area. The bulls were slaughtered based on the usual practices of the Brazilian beef industry according to the Brazilian RIISPOA—Regulation of Industrial and Sanitary Inspection of Animal Products. The carcasses were then divided medially from the sternum to the spine, resulting in two similar halves. After these procedures, muscle samples were collected from the LT muscle of the left half-carcass at the 12th to 13th rib. The muscle samples were frozen in liquid nitrogen and stored at −80 °C at the laboratory for gene expression and enzyme analyses. Subsequently, the half carcasses were washed, identified, and stored in a chilling chamber at 4 °C for 24 h. After chilling, the LT muscle samples were collected from the left side of the carcasses between the 12th and 13th ribs and stored at −20 °C for beef chemical composition and fatty acid analysis. The samples were transported to the Animal Science Laboratory at São Paulo State University (Jaboticabal, Brazil).

2.5. Chemical Composition of Beef

To determine the chemical composition of beef, the steaks were thawed at 4 °C for 24 h, ground, and subjected to composition analyses (crude fat, ash, crude protein, and moisture) using the FoodScan Meat Analyser™® (FOSS, Hillerod, Denmark) with a near-infrared spectrophotometer (analyses AOAC method: 2007-04).

2.6. Fatty Acid Profile of Beef and Diet

Sample lipids were extracted according to Bligh and Dyer [14]. Briefly, 15 g of meat sample was subjected to extraction with a chloroform–methanol mixture (2:1 ratio) and then transmethylated [15]. A 1 μL aliquot of transmethylated lipid was injected into a gas chromatograph (Shimadzu GC-2010 Plus; Shimadzu Corporation, Kyoto, Japan) with a flame ionisation detector and capillary column (Restek-RT® 2560, Bellefonte, PA, USA; 100 m long, 0.25 mm internal diameter, and 0.20 μm film thickness). Hydrogen was used as the carrier gas at a flow rate of 1.0 mL/min. The temperature program of the oven of the gas chromatograph began at 100 °C with a standby time of 5 min and then increased to 240 °C (4 °C/min) with a standby time of 20 min. The detector temperature was 260 °C. Identification and quantification of the proportion of fatty acids were performed by comparing their retention times with those of commercial standards of total fatty acid methyl esters (Supelco 37 component FAME mix; conjugated linoleic acid methyl ester (trans10–cis12) and conjugated linoleic acid methyl ester (cis9–trans11); Sigma-Aldrich, Bellefonte, PA, USA). The results are expressed as mg/100 g of beef.

2.7. Lipogenic Enzyme Activity

Approximately 1.5 g of LT was cut and placed in 4.5 mL of 0.1 M phosphate buffer (K_2HPO_4, pH 7.4, 25 °C), homogenised, and centrifuged at 3000× *g* for 15 min at 4 °C. The pellet was then discarded, and the supernatant was centrifuged at 15,000× *g* for 30 min at 4 °C. The resulting supernatant fractions were used for enzyme measurements. NADP-malate dehydrogenase and isocitrate dehydrogenase enzyme activity were measured as described by Martin [16] and Bergmeyer and Bernt [17,18], respectively. All enzyme assays were performed in duplicate using the spectrophotometric absorbance of the solutions in cuvettes at 340 nm. The slopes of the linear rates of NADPH production were used to calculate enzyme activity.

2.8. Gene Expression Analyses

The target and reference primers were designed using sequences registered and published in the GenBank public data bank of the National Center for Biotechnology Information platform (Table 2). Primers were designed using OligoPerfect Designer software (Invitrogen, Karlsruhe, Germany) and synthesized (Invitrogen, Carlsbad, CA, USA). Nine target genes (*PPARG, SREBP-1c, SCD1, ACCA, LPL, FABP4, ACOX, CPT2,* and *PPARA*) and two target reference genes (*β-actin* and glyceraldehyde-3-phosphate dehydrogenase (*GAPDH*)) were used, as proposed by Vandesompele et al. [19]. Total RNA extraction was performed using a RNeasy Lipid Tissue Mini Kit (Qiagen, Valencia, CA, USA). RNA contamination (260/280 and 260/230) and concentration (ng/µL) were assessed using a NanoDrop 1000 Spectrophotometer (Thermo Fisher Scientific, Santa Clara, CA, USA, 2007). RNA quality was assessed using an Agilent 2100 Bioanalyzer (Agilent, Santa Clara, CA, EUA, 2009) and the Agilent RNA 6000 Nano Chip kit (Agilent, Santa Clara, CA, USA).

Table 2. Primer sequences used for quantitative RT-PCR analyses.

Gene Abbrevation	Gene	Primer	R^2	Efficiency
PPARG	*Peroxisome proliferator-activated receptor gamma*	F: CGATATCGACCAACTGAACC R: AACGGTGATTTGTCTGTCGT	0.992	90.788
SREBP-1c	*Sterol regulatory element-binding protein-1c*	F: GAGCCACCCTTCAACGAA R: TGTCTTCTATGTCGGTCAGCA	0.999	100.593
SCD1	*Stearoyl-CoA desaturase*	F: TTATTCCGTTATGCCCTTGG R: TTGTCATAAGGGCGGTATCC	0.997	94.776
ACACA	*Acetyl CoA carboxylase alfa*	F: TGAAGAAGCAATGGATGAACC R: TTCAGACACGGAGCCAATAA	0.998	101.32
LPL	*Lipoprotein lipase*	F: CTCAGGACTCCCGAAGACAC R: GTTTTGCTGCTGTGGTTGAA	0.993	94.257
FABP4	*Fatty acid binding protein 4*	F: GGATGATAAGATGGTGCTGGA R: ATCCCTTGGCTTATGCTCTCT	0.997	90.259
ACOX	*Acyl-CoA oxidase 1*	F: GCTGTCCTAAGGCGTTTGTG R: ATGATGCTCCCCTGAAGAAA	0.991	90.993
CPT2	*Carnitine Palmitoyltransferase 2*	F: CATGACTGTCTCTGCCATCC R: ATCACTTTTGGCAGGGTTCA	0.991	94.577
PPARA	*Peroxisome proliferator-activated receptor alfa*	F: CAATGGAGATGGTGGACACA R: TTGTAGGAAGTCTGCCGAGAG	0.994	91.665
β-Actin	*β-actin*	F: GTCCACCTTCCAGCAGATGT R: CAGTCCGCCTAGAAGCATTT	0.998	93.059
GAPDH	*Glyceraldehyde 3 phosphate*	F: CGACTTCAACAGCGACACTC R: TTGTCGTACCAGGAAATGAGC	0.994	92.896

The cDNA synthesis was performed using SuperScript III First-Strand Synthesis SuperMix for qRT-PCR (Invitrogen, Carlsbad, CA, USA), according to the manufacturer's instructions. A 7500 Real-Time PCR system (Applied Biosystems, Foster, CA, USA, 2009) was used for qPCR with a SYBR Green RT-PCR kit from Bio-Rad. The cycling conditions were 2 min polymerase activation at 95 °C and 40 cycles at 95 °C for 15 s and 60 °C for 30 s. A validation assay of amplification efficiencies demonstrated that the target and reference genes were approximately equivalent. Relative mRNA expression was calculated according to $\Delta C_T = C_T$ (target gene) $- C_T$ (average reference genes). The calibration was determined using the formula $\Delta\Delta C_T = \Delta C_T$ (sample) $- \Delta C_T$ (calibrator), and the mineral finished in pasture plus supplementation treatment was used for each breed. Relative expression was evaluated using the $2^{-\Delta\Delta CT}$ formula [20].

2.9. Gene Set Enrichment Analysis

The ClueGO of the Cytoscape program 3.7.1 was used for the enrichment analysis with the genes studied using the bovine genome UMD 3.1. An enrichment analysis was performed to visualise non-redundant biological terms for genes using the ClueGO plug-in of Cytoscape [21], with bovine genome UMD 3.1 (http://www.ncbi.nlm.nih.gov/genome/?term=bos+taurus, accessed on 22 August 2019) as reference. The genes were enriched considering the gene ontology (GO) biological processes classification system.

2.10. Statistical Analysis

The experimental design was a completely randomised block (by weight: light and heavy; and by breed: Nellore, ½ Angus, and ½ Senepol) in a 2 × 2 factorial arrangement, with two supplements administered during the growing (mineral or protein + energy supplementation) and two finishing systems (pasture and feedlot systems). All data were analysed using PROC MIXED software of SAS 9.4 (SAS Inst. Inc., Cary, NC, USA, Cary Inc., Rural Hall, NC, USA). The statistical model included treatments and all interactions as fixed effects and bulls nested within the paddock and breed as random effects. The mean and standard error of the mean were calculated for each variable (the experimental unit was the animal; $n = 30$/ treatment). When significant main or interaction effects were detected, Tukey's test ($p \leq 0.05$) was used to determine the differences between means.

3. Results

3.1. Meat Composition

The feed strategy in the growing (mineral vs. protein + energy supplement) did not result in higher lipid content in beef ($p \geq 0.05$). At the end of the finishing phase, the cattle fed with mineral or protein + energy had lipid averages of 2.03 and 2.19 g/100 g of beef, respectively (Table 3). The finishing system affected the lipid content ($p < 0.001$) and moisture ($p < 0.001$; Table 3). The meat of bulls finished in the feedlot system had an 86.48% higher lipid concentration than meat from bulls finished in the pasture system (2.75 vs. 1.48 g/100 g of beef, respectively). The increase in lipid content resulted in a decrease ($p \leq 0.001$) in the moisture content of beef from the feedlot group compared to beef from the pasture group, with averages of 72.55 vs. 73.74 g/100 g, respectively. The animals presented beef with similar ash and protein contents ($p > 0.050$; Table 3). No interaction between the growing feed and finishing system was observed ($p > 0.050$) for the chemical composition.

Table 3. Hot carcass weight and chemical composition (lipid, ash, protein and moisture [g/100 g of meat]) of *Longissimus thoracis* meat from young bulls supplemented with mineral or protein + energy during the growing phase and finished in pasture plus concentrate or feedlot system.

Finishing System	Feedlot [1]		Pasture [2]		SEM [5]	GF [6]	FS [7]	GF × FS
Growing Feed	MIN [3]	PRE [4]	MIN [3]	PRE [4]				
Hot carcass weight, kg	297	297	272	285	13.10	0.124	<0.001	0.096
Lipid	2.66	2.84	1.41	1.54	0.180	0.466	<0.001	0.868
Ash	2.22	2.48	2.64	2.53	0.128	0.623	0.071	0.150
Protein	22.3	22.4	22.0	22.4	0.199	0.255	0.370	0.424
Moisture	72.8	72.3	73.9	73.6	0.296	0.129	<0.001	0.826

[1] Feedlot (75:25 corn silage:concentrate ratio); [2] Pasture (20 g/kg BW/day of concentrate); [3] Mineral (ad libitum); [4] Protein + energy (3 g/kg BW/day); [5] Standard error of mean; [6] Growing feed; [7] Finishing system.

3.2. Fatty Acid Profile and Enzyme

In the current study, the total saturated fatty acid (SFA) profile was not affected ($p > 0.050$) during the growing feed (Table 4). A higher total concentration of SFA ($p = 0.003$) was observed in the meat of bulls finished in the feedlot system, which was due to the increase ($p \leq 0.003$) in myristic (C14:0), palmitic (C16:0), and margaric (C17:0) fatty acids found in this group (Table 4).

Table 4. Effect of mineral or protein + energy supplement during the growing phase and finishing with intensive supplementation in pasture plus concentrate or feedlot system on fatty acid profile (mg/100 g of meat) from *Longissimus thoracis* of young bulls.

Finishing System	Feedlot [1]		Pasture [2]		SEM [5]	GF [6]	FS [7]	GF × FS
Growing Feed	**MIN [3]**	**PRE [4]**	**MIN [3]**	**PRE [4]**				
C12:0	0.92	0.64	0.62	0.67	0.108	0.339	0.241	0.064
C14:0	62.90	45.04	32.14	34.80	6.824	0.300	0.001	0.060
C15:0	4.94	4.21	4.04	4.26	0.563	0.854	0.440	0.211
C16:0	614.60	495.97	352.85	354.25	61.644	0.453	0.000	0.249
C17:0	19.69	15.69	12.33	12.46	2.202	0.432	0.003	0.222
C18:0	424.43	364.77	306.54	324.67	49.170	0.948	0.068	0.338
C14:1	14.27	9.30	8.87	8.51	1.840	0.136	0.056	0.110
C15:1	2.20	1.84	1.88	1.99	0.265	0.995	0.959	0.275
C16:1	65.54	53.00	44.79	41.72	7.339	0.347	0.010	0.373
C17:1	10.38	9.07	7.81	7.17	1.026	0.307	0.005	0.533
C18:1n9c	857.23	798.07	584.83	582.47	85.856	0.873	0.000	0.491
C18:2c9–t11	62.24	32.73	25.67	47.44	15.630	0.678	0.366	0.021
C18:2t10–c12	14.93	13.26	15.55	14.37	1.247	0.231	0.267	0.694
C18:2n6c	107.57	107.58	131.22	124.87	10.954	0.838	0.006	0.555
C18:3n6	1.05	0.95	1.09	1.00	0.079	0.192	0.334	0.972
C18:3n3	6.11	6.53	7.90	7.22	0.606	0.675	0.001	0.261
C20:2n6	1.67	1.70	1.46	1.38	0.124	0.957	0.005	0.727
C20:3n6	4.11	4.30	5.75	5.01	0.545	0.504	0.003	0.182
C20:3n3	0.32	0.25	0.24	0.27	0.040	0.486	0.481	0.102
C20:4n6	19.00	21.33	27.77	25.75	2.560	0.922	0.000	0.122
C20:5n3	2.45	2.52	2.70	2.74	0.229	0.777	0.164	0.778
C22:6n3	1.35	0.60	0.42	0.62	0.346	0.509	0.133	0.119
∑SFA [8]	1129.55	926.32	708.52	740.42	117.232	0.648	0.003	0.228
∑MUFA [9]	936.63	871.28	648.18	641.85	93.766	0.849	0.001	0.502
∑PUFA [10]	239.88	194.06	220.18	230.67	18.862	0.245	0.382	0.060
∑n6–PUFA [11]	133.65	135.86	167.30	158.02	13.799	0.866	0.003	0.425
∑n3–PUFA [12]	9.86	10.09	11.08	10.85	0.783	0.993	0.090	0.673
n6:n3	12.02	13.42	14.55	14.70	0.929	0.286	0.009	0.312
PUFA:SFA	0.24	0.23	0.33	0.32	0.034	0.403	0.001	0.997

[1] Feedlot (75:25 corn silage:concentrate ratio); [2] Pasture (20 g/kg BW/day of concentrate); [3] Mineral (ad libitum); [4] Protein + energy (3 g/kg BW/day); [5] Standard error of mean; [6] Growing feed; [7] Finishing phase; [8] Sum of 12:0, 14:0, 15:0, 16:0, 17:0 and 18:0; [9] Sum of 14:1, 15:1, 16:1, 17:1 and 18:1; [10] Sum of 18:2 c9–t11, 18:2 t10–c12, 18:2n6, 18:3n6, 18:3n3, 20:2n6, 20:3n6, 20:3n3, 20:4n6, 20:5n3 and 22:6n3; [11] Sum of 18:2n6, 18:3n6, 20:2n6, 20:3n6 and 20:4n6; [12] Sum of 18:3n3, 20:3n3, 20:5n3 and 22:6n3.

The total MUFA (Table 4) increased in the meat of bulls from the feedlot system compared to the pasture system ($p = 0.001$; 903.95 vs. 645.01 mg/100 g beef, respectively). In addition, increases ($p \leq 0.010$) in the palmitoleic (C16:1), heptadecenoic (C17:1), and oleic (C18:1n9c) acid content were observed in the meat of the bulls from the feedlot compared to the pasture finishing system (59.27 vs. 43.26, 9.72 vs. 7.49, and 827.65 vs. 583.65, mg/100 g beef, respectively; Table 4). In the current study, the pasture group presented higher ($p \leq 0.003$) arachidonic (C20:4n6) and dihomo-γ-linolenic (C20:3n6) fatty acid levels than the feedlot system (26.76 vs. 20.17; 5.38 vs. 4.21 mg/100 g beef, respectively; Table 4). The meat of the bulls from the pasture system had ($p \leq 0.006$) higher linoleic (C18:2n6) and α-linolenic (C18:3n3) fatty acid concentration (128.04 vs. 107.57; 7.56 vs. 6.32 mg/100 g beef, respectively; Table 4).

A similar concentration in C18:2 trans10–cis12 fatty acid was observed between treatments ($p > 0.050$). The pasture system with intensive supplementation increased in n6 total and n6:n3 ratio ($p \leq 0.009$; 162.66 vs. 134.76; 14.63 vs. 12.72 mg/100 g meat, respectively) and tended ($p = 0.090$; 10.97 vs. 9.98 mg/100 g meat, respectively) to provide greater n3 total concentrations in beef compared to the feedlot system. An interaction ($p = 0.021$) was observed between the growing feed and finishing system for C18:2 cis9–trans11.

The growing feed or finishing system did not alter ($p \geq 0.050$) the NADP–Malate dehydrogenase enzyme activity. Isocitrate dehydrogenase enzyme activity was lower ($p = 0.020$) in the meat of bulls from feedlot than in the pasture finishing system (2788.07 vs. 3279.54 nmol/min, respectively; Table 5).

Table 5. NADP–malate dehydrogenase and isocitrate dehydrogenase (nmol/min) in the *Longissimus thoracis* muscle from young bulls fed mineral or protein + energy supplement during the growing phase and finished in pasture or feedlot.

Finishing System	Feedlot [3]		Pasture [4]		SEM [7]	GF [8]	FS [9]	GF × FS
Growing Feed	**MIN [5]**	**PRE [6]**	**MIN [5]**	**PRE [6]**				
Isocitrate [1]	2645.6	2930.5	3520.4	3038.7	245.0	0.820	0.020	0.066
NADP-Malate [2]	47.2	46.5	48.6	43.6	3.8	0.472	0.888	0.554

[1] Isocitrate dehydrogenase; [2] NADP–Malate dehydrogenase; [3] Feedlot (75:25 corn silage:concentrate ratio); [4] Pasture (20 g/kg BW/day of concentrate); [5] Mineral (ad libitum); [6] Protein + energy (3 g/kg BW/day); [7] Standard error of mean; [8] Growing feed; [9] Finishing system.

3.3. Gene Expression

The muscle of bulls fed the feedlot system had greater ($p < 0.001$) expression levels of *stearoyl-CoA desaturase* (*SCD1*) compared to the muscle of bulls from the pasture system (Figure 1). In addition, the animals fed protein + energy during the growing presented higher ($p = 0.020$) *SCD1* expression compared to animals that received mineral supplementation.

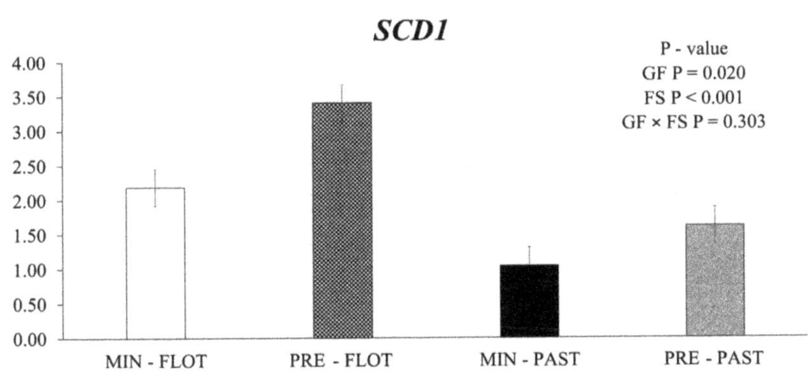

Figure 1. Relative expression of *SCD1* (*stearoyl-CoA desaturase*) in the *Longissimus thoracis* muscle from young bulls fed mineral (ad libitum; MIN) or protein + energy (3 g/kg BW/day; PRE) during the growing and finished in pasture (20 g/kg BW/day of concentrate; PAST) or feedlot (75:25 corn silage:concentrate ratio; FLOT). GP = Growing feed; FS = Finishing system.

An interaction among growing feed and finishing system ($p = 0.026$) was observed, with *sterol regulatory element-binding protein-1c* (*SREBP1c*) having lower mRNA expression in the muscle of bulls finished in the pasture system regardless of the supplementation during the growing phase (Figure 2A). In addition, lower ($p = 0.013$) *peroxisome proliferator-activated receptor gamma* (*PPARG*) mRNA expression was detected in the muscle of bulls fed mineral during the growing phase and finished in the pasture system (Figure 2B).

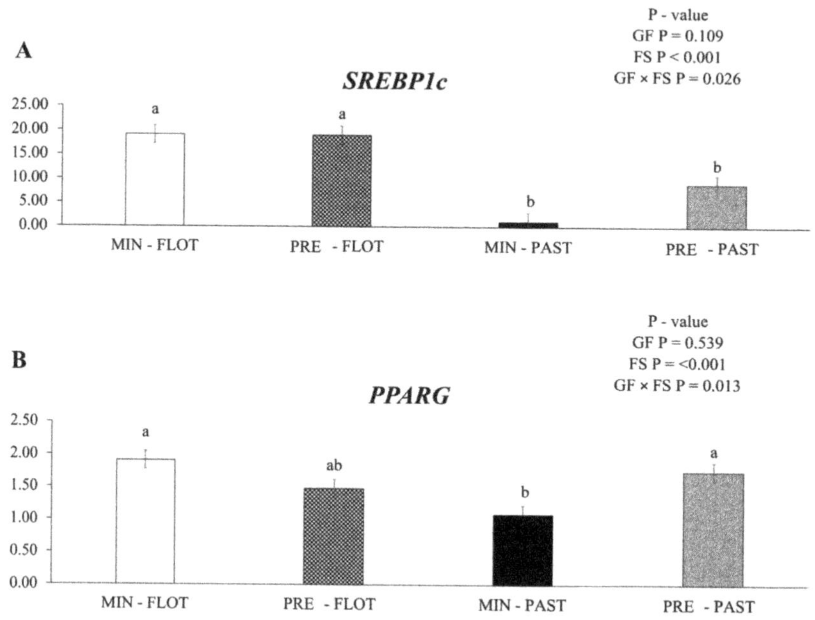

Figure 2. Relative expression of *SREBP-1c* (*sterol regulatory element-binding protein-1c*); (**A**) and *PPARG* (*peroxisome proliferator-activated receptor gamma*); (**B**) in the *Longissimus thoracis* muscle from young bulls fed mineral (ad libitum; MIN) or protein + energy (3 g/kg BW/day; PRE) during the growing and finished in pasture (20 g/kg BW/day of concentrate; PAST) or feedlot (75:25 corn silage:concentrate ratio; FLOT). GP = Growing feed; FS = Finishing system. The means without a common letter are different ($p < 0.050$).

The expression of *acetyl CoA carboxylase alfa* (*ACACA*) mRNA was upregulated ($p = 0.025$; Figure 3A) in the muscle of bulls fed protein + energy compared to mineral feed, regardless of the finishing system (1.51 vs. 1.19). An increase in *ACACA* mRNA expression in the muscle of bulls from the feedlot finishing system compared to the pasture finishing system was not observed ($p = 0.149$).

In the current study, an interaction ($p \leq 0.001$) was found between the growing feed and finishing system for *fatty acid binding protein 4* (*FABP4*) mRNA (Figure 3B). Greater expression was detected in the muscle of bulls fed mineral during the growing and finished in the feedlot system. Moreover, bulls fed with protein + energy supplement followed by the feedlot system had intermediate values, and the two lowest values were found in the muscle of bulls finished in the pasture system regardless of the feed during the growing (mineral or protein + energy) (4.71, 3.47, 1.06, and 1.41, respectively).

An interaction ($p \leq 0.001$) between the growing feed and finishing systems was observed for the *lipoprotein lipase* (*LPL*) mRNA expression (Figure 3C). The muscle of bulls from protein + energy during the growing and finished in the feedlot system had higher levels of *LPL* mRNA expression than the muscle of bulls fed mineral during the growing and finished in the feedlot system.

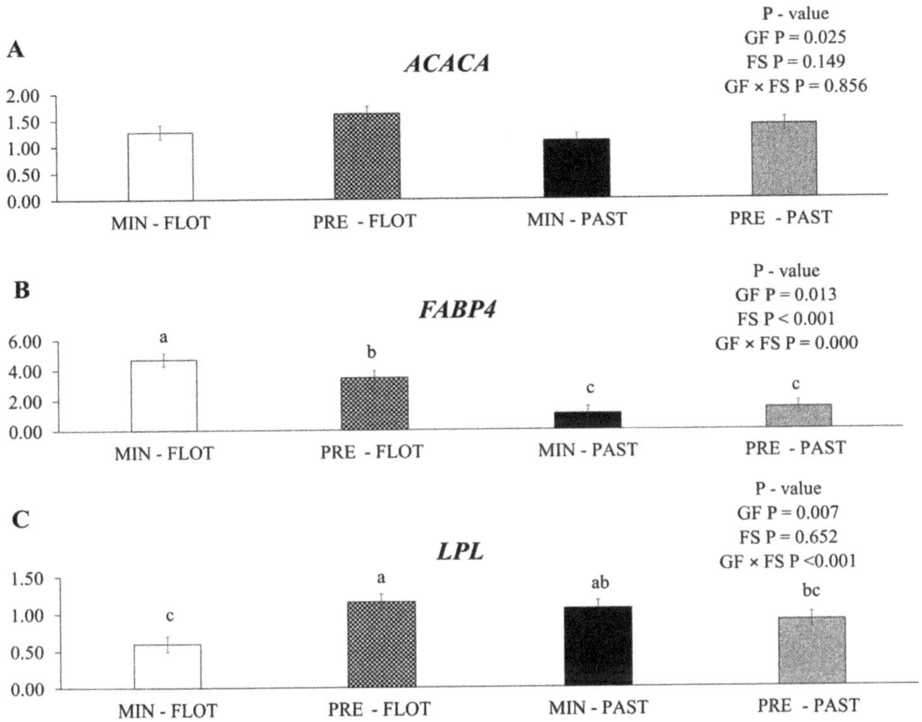

Figure 3. Relative expression of *ACACA* (*acetyl CoA carboxylase alfa*); (**A**), *FABP4* (*fatty acid binding protein 4*); (**B**) and *LPL* (*lipoprotein lipase*); (**C**) in the *Longissimus thoracis* muscle from young bulls fed mineral (ad libitum; MIN) or protein + energy (3 g/kg BW/day; PRE) during the growing and finished in pasture (20 g/kg BW/day of concentrate; PAST) or feedlot (75:25 corn silage:concentrate ratio; FLOT). GP = Growing feed; FS = finishing system. The means without a common letter are different ($p < 0.050$).

In this study, an interaction ($p = 0.013$) was found in the muscle of bulls between the growing and finishing systems (Figure 4A) for *peroxisome proliferator-activated receptor alfa* (*PPARA*) mRNA expression, where this gene was upregulated in the muscle of bulls fed with mineral during the growing and finished in the pasture system.

An interaction between the growing feed and finishing systems was observed for the gene encoding *carnitine palmitoyl transferase 2* (*CPT2*; $p < 0.001$; Figure 4B). The muscle of bulls fed protein + energy during the growing and finished in the feedlot system had higher mRNA expression of the *CTP2* gene than the muscle of bulls fed protein + energy and finished in the pasture system. In contrast, the muscle of bulls fed mineral and finished in the feedlot or pasture system had intermediate *CPT2* mRNA expression (1.24, 0.68, 0.79, and 1.05, respectively). No treatment effects ($p \geq 0.050$) or interactions were found for *acyl CoA oxidase 1* (*ACOX*, Figure 4C).

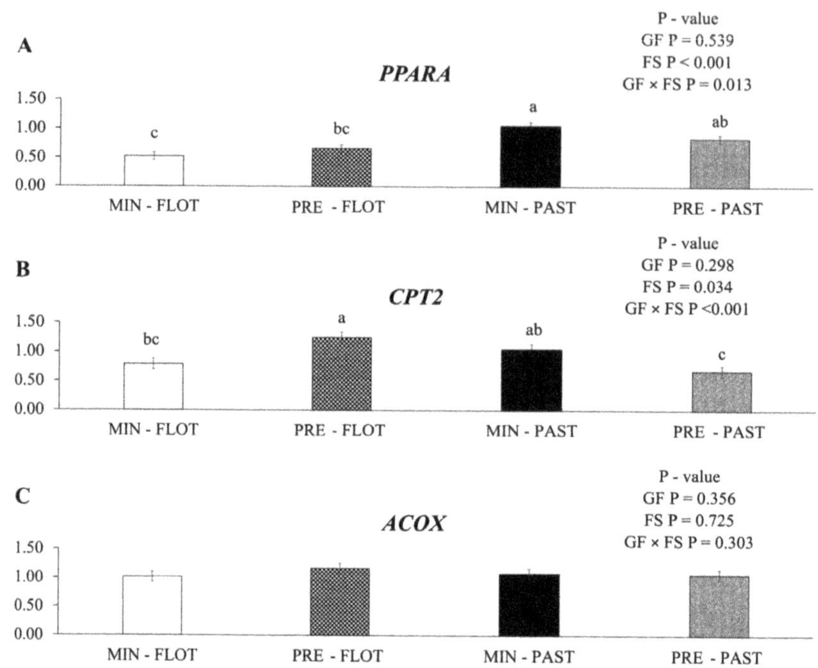

Figure 4. Relative expression of *PPARA* (*peroxisome proliferator-activated receptor alfa*); (**A**), *CPT2* (*carnitine palmitoyl transferase* 2); (**B**) and *ACOX* (*acyl CoA oxidase 1*); (**C**), in the *Longissimus thoracis* muscle from young bulls fed mineral (ad libitum; MIN) or protein + energy (3 g/kg BW/day; PRE) during the growing phase and finished in pasture (20 g/kg BW/day of concentrate; PAST) or feedlot (75:25 corn silage:concentrate ratio; FLOT). GP = Growing feed; FS = Finishing system. The means without a common letter are different ($p < 0.050$).

4. Discussion

To the best of our knowledge, this is the first study to investigate the relative expression of genes related to lipid metabolism in the intramuscular adipose tissue of LT muscle from cattle supplemented during the growing phase and finished in the tropical pasture or feedlot with intensive supplementation. We hypothesized that the dietary treatment with protein + energy during the growing followed by the feedlot finishing system could increase lipogenic genes and decrease lipolytic genes, resulting in increased lipid content in the meat of bulls. However, our hypothesis was not confirmed, which can be explained by the hot carcass weight (HCW) and the degree of marbling score [22] that the animals reached. At the end of the growing phase, the animals from the protein + energy group presented 21.9 kg body weight more than the mineral group (data not presented). However, this difference was diluted during the finishing phase (120 days), where the bulls had a similar value of HCW (284 and 291 kg for protein + energy and Mineral, respectively).

The bulls from the pasture system have more significant energy expenditure when compared to the feedlot system, which reduces energy available required for lipid deposition in meat. Animals in the pasture system have a higher energy expenditure in response to more activity related to feeding even when they are administered the concentrate [23]. Overall, the bulls require greater movement (physical activity), collection, and selection of pasture (source of roughage), which promotes greater expenditure and energy requirement than those in the feedlot [8]. In addition, animals on pasture showed an increase of 11% in metabolic energy requirements for maintenance when compared to the animals in the feedlot system [24].

The fatty acid profile was examined in the sample collected at the end of the finishing phase, which had a 129-day duration, long enough to change the fatty acid profile. Lipogenesis occurs through dietary lipid absorption and de novo fatty acid synthesis in animals [5]. The increase in these fatty acids in beef from the feedlot group, where the corn silage diet had lower total SFA than the pasture composition (Table 1), could be due to an increase in de novo fatty acid synthesis and higher enzyme activity of fatty acid synthase, a key enzyme in the lipogenic pathway that catalyses the reactions of fatty acid biosynthesis and conversion of acetyl-CoA and malonyl-CoA to palmitic acid, which may have increased the isomers of SFA [25].

Isocitrate dehydrogenase plays a crucial role in lipid metabolism, catalysing the conversion of oxidative decarboxylation of isocitrate to α-ketoglutarate with the production of NADPH [26] for de novo fatty acid synthesis. The higher values found in isocitrate dehydrogenase enzyme activity suggest that the feedlot system can provide higher energy, as this enzyme can be reduced by increasing the energy levels [27].

The high concentrations of oleic and palmitoleic acid in the meat of bulls from the feedlot system are related to *SCD1* expression (Δ9 desaturase enzyme). *SCD1* is associated with the biosynthesis of unsaturated fatty acids (Figure 5) and is a key enzyme that catalyses the desaturation of a range of fatty acyl-CoA substrates, mainly palmitoyl and stearoyl, resulting in palmitoleic and oleic acid, respectively [28]. According to Smith et al. [29], the accumulation of MUFA in adipose tissues coincides with an increase in *SCD1* gene expression. The increase in monounsaturated acid, such as oleic acid, is related to meat palatability [30], and oleic fatty acid represents the largest amount of monounsaturated acid in beef [31], ultimately aligning with our results, regardless of the treatments.

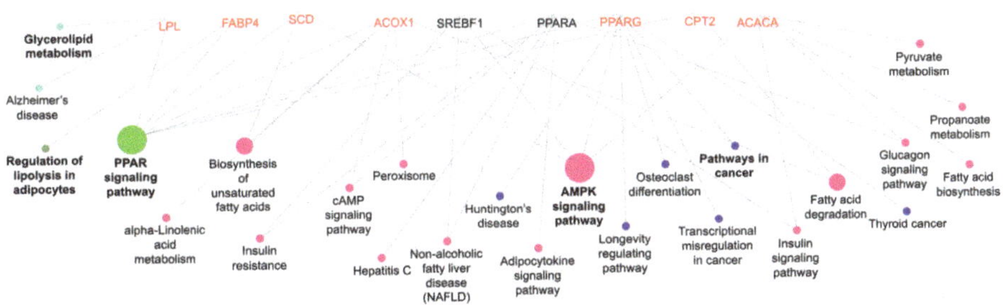

Figure 5. Functional classification of the lipid metabolism genes for the biological process GO category. The edges represent the interaction between the genes and processes. The node fill color represents the relationship of genes caused by a common process (i.e., purple nodes represent the process associated with two or more genes and green node represents the process with the most relationship). The genes are grouped by a common process.

The meat of cattle fed exclusively with grass or grass plus supplements had greater α-linolenic fatty acid content [32], while the beef of animals fed grain had higher linoleic content due to diet composition. In the current study, the pasture group presented higher linolenic fatty acid levels and, interestingly, higher linoleic concentration than the feedlot system, which was not expected, suggesting that the bulls finished in the feedlot system could have higher ruminal biohydrogenation activity, leading to a decrease in linoleic acid in beef. Bulls finished in the pasture system had a higher decline and variation in ruminal pH once the concentrate supplementation was offered separately from roughage, which may have resulted in a reduction in microorganisms that play an important role in biohydrogenation [33].

γ-Linolenic acid, eicosatrienoic acid (C20:2n6), and arachidonic acid are produced from linoleic acid by the action of desaturase and elongase enzymes, whereas eicosapentaenoic acid (C20:5n3) and docosahexaenoic acid (C22:6n3) are produced from alpha-linolenic acid [34]. In this context, the increase in linoleic acid in the meat of bulls from the pasture system aligned with the increase in arachidonic and dihomo-γ-linolenic fatty acids, which was not expected but could have caused an increase in n6:n3 ratio values due to a higher n6 total concentration in this group. The lower n6:n3 ratio is recommended for the benefits of human health; it is found to be around 1 for beef produced using the grass diet [35–37]. The n6:n3 ratio in beef increases with the increased inclusion of concentrate supplements in grazing animal diets [38]. Although some studies show that concentrate supply has not changed n6:n3 in the meat, these works have tested concentrate inclusion in up to 50% of the diet. In this sense, in the present study, this change may be related to the high supply of concentrate (2% BW) for animals in the pasture system.

The C18:2 trans10–cis12 can be produced because of rumen pH reduction as a grain-based diet and can decrease the relative abundance of *SREBP1c*, a gene responsible for encoding sterol regulatory element-binding protein [38,39], consequently contributing to a reduction in fat deposition. However, this result was not found in the current study, where C18:2 trans10–cis12 in the meat of bulls had a similar concentration, which means that it was not related to changes in *SREBP1c* activity.

In this way, some n3 fatty acid isomers, such as arachidonic and docosahexaenoic acid, can affect the expression of genes related to lipid metabolism by controlling *SREBP1c*, which is the main gene controlling lipogenesis [40]. Therefore, the pasture system with intensive supplementation tended to provide greater n3 concentrations in beef, which may have helped to decrease intramuscular fat concentrations.

The expression levels of *SREBP1c* are related to energy availability, and it is a major factor in the expression of genes related to fat deposition [41]. Although not measured in this study, the mineral feed during the growing phase followed by the pasture finishing system may have provided lower blood insulin and glucose concentrations. Accordingly, the decrease in insulin may have decreased the concentration of *SREBP1c*, as this gene is associated with two pathways (Figure 5), insulin signalling and AMPK signalling. Such findings indicate that insulin can control *SREBP1c* abundance and induce de novo lipogenesis [42] and agree with the results of a higher concentration of isocitrate dehydrogenase enzyme activity in the mineral feed during the growing followed by the pasture finishing system (3520.39 nmol/min). Moreover, such results also help explain the *SCD1* gene expression in our study, considering that *SCD1* activity is increased by dietary glucose, fructose, and insulin [43].

In addition, the above explanation can be related to lower levels of *PPARG* expression in the muscle of bulls fed mineral during the growing phase and finished in the pasture system. *PPARG* can regulate biological processes, such as lipid metabolism; however, it is more highly expressed in adipose tissue and participates in adipogenesis and insulin sensitivity [44]. The *PPARG* is linked to *SREBP1c* and *ACACA* by the AMPK signalling pathways (Figure 5).

ACACA is associated with the biosynthesis of fatty acids and it is involved in the first step of SFA synthesis and the enzyme carboxylation of acetyl-CoA into malonyl-CoA in response to diet and hormones [45]. An increase in *ACACA* mRNA expression in the muscle of bulls from the feedlot group was expected but was not observed. Thus, the possible increase in SFA by de novo fatty acid synthesis in the feedlot group may occur due to the increased activity of other enzymes, such as fatty acid synthase; however, this enzyme was not measured in this study. The expression levels of *ACACA* and fatty acid synthase are related to fatty acid biosynthesis and are not regulated in coordination [38].

The transport of fatty acids into cells is facilitated by *FABP4* [46]; thus, the increase in its gene expression can be related to an increase in triacylglycerols from diets. The lowest *FABP4* expression in the muscle of bulls finished in the pasture group, regardless of the growing supplementation, may have been due to the lower intramuscular fat of this group. Yang et al. [47] found an increase in *FABP4* levels when intramuscular fat was increased by a higher-energy diet compared to a lower-energy diet.

Although *LPL* and *FABP4* can exhibit complementary functions [38], there is no similar relationship between the expression of both genes due to sampling of both adipocytes, which may express more FABP4, and myocytes, which may express more *LPL*. Lipoprotein lipase is an enzyme that catalyses the hydrolysis of triglycerides present in lipoproteins [47]; thus, the muscle of bulls fed protein + energy and finished in the feedlot system may have higher lipid turnover and higher triglyceride hydrolysis into non-esterified fatty acids to supply energy for tissues, which leads to increased gene expression.

Although *PPARA* is highly expressed in the liver [48], this enzyme is responsible for peroxisome proliferator-induced responses, including the transcriptional activation of genes involved in fatty acid oxidation [44]. In this study, the less *PPARA* expression in the muscle of bulls fed with mineral supplementation during the growing followed by the pasture finishing system, suggesting that the muscle of these bulls had a higher lipid oxidation rate. The *CPT2* and *ACOX1* are other genes that may participate in fatty acid oxidation mechanisms in muscle and are related to the fatty acid degradation pathway (Figure 5); however, no change was observed in *ACOX1* mRNA expression. In this study, higher lipid content was expected in meat from animals fed protein + energy during the growing and finished in the feedlot; however, the up-regulated expression of *CPT2* in these animals may have influenced the degree of marbling, considering the role of this enzyme in mitochondrial long-chain fatty acid oxidation [49].

Fat deposition occurs in response to the activation of metabolic actions, such as lipogenesis and lipolysis, which occur according to the need for lipid release or storage, as well as the interaction of dietary energy and the level of expression of the gene's relationship with lipids [41]. Accordingly, even if the hypothesis of the study was not confirmed, the supplementation strategy during the growing phase affected gene expression but did not result in higher lipid content at slaughter.

5. Conclusions

During the growing phase, the supplementation strategy of bulls showed changes in the lipolytic and lipogenic genes, but not enough to cause changes in intramuscular fat at the finishing stage. The finishing system impacted the genes and affected the intramuscular fat and fatty acid. In addition, our results indicate that beef from pasture-fed bulls supplemented with concentrate had a greater concentration of total saturated and a higher n6:n3 ratio in meat, which is considered non-ideal for human health.

Author Contributions: Conceptualization, J.A.T., T.T.B., E.S.V. and J.M.D.; methodology, J.A.T. and L.S.F.; validation, J.A.T. and T.A.S.; formal analysis, J.A.T.; investigation, J.A.T.; data curation, J.A.T.; writing—original draft preparation, J.A.T., E.S.V. and G.F.; writing—review and editing, J.A.T., G.D.R.-Z., R.d.N.S.T. and G.L.P.; visualization, G.D.R.-Z., R.d.N.S.T., R.A.C., W.A.B., L.A.L.C. and O.R.M.N.; supervision, E.S.V.; project administration, T.T.B. All authors have read and agreed to the published version of the manuscript.

Funding: This research was funded by the Research Foundation of the State of São Paulo (Fundação de Amparo à Pesquisa do Estado de São Paulo—FAPESP; grant number 2015/01147-0). This study was also financed in part by CAPES (Coordenação de Aperfeiçoamento de Pessoal de Nível Superior), Finance code 001, and the FUNEP (Fundação de Apoio a Pesquisa, Ensino e Extensão).

Institutional Review Board Statement: The study was conducted in accordance with the Animal Use Ethics Committee of the School of Agriculture and Veterinary Sciences (FCAV), UNESP, Jaboticabal, São Paulo, Brazil (certificate number 5628/15).

Informed Consent Statement: Not applicable.

Data Availability Statement: All relevant data are presented within the paper.

Acknowledgments: The authors thank to the Fundação de Amparo à Pesquisa do Estado de São Paulo—FAPESP (grant: 2015/01147-0) for the financial and resources support, including the analysis and samples required for the execution of this research, J.A.T. (process number 2016/10953-3) and the Coordenação de Aperfeiçoamento de Pessoal de Nível Superior—CAPES.

Conflicts of Interest: The authors declare no conflict of interests.

References

1. Owens, F.N.; Dubeski, P.; Hanson, C.F. Factors that alter the growth and development of ruminants. *J. Anim. Sci.* **1993**, *71*, 3138–3150. [CrossRef] [PubMed]
2. Roth, M.T.P.; Resende, F.D.; Oliveira, I.M.; Fernandes, R.M.; Custódio, L.; Siqueira, G.R. Does supplementation during previous phase influence performance during the growing and finishing phase in Nellore cattle? *Livest. Sci.* **2017**, *204*, 122–128. [CrossRef]
3. Sampaio, R.L.; de Resende, F.D.; Reis, R.A.; de Oliveira, I.M.; Custódio, L.; Fernandes, R.M.; Pazdiora, R.D.; Siqueira, G.R. The nutritional interrelationship between the growing and finishing phases in crossbred cattle raised in a tropical system. *Trop. Anim. Health Prod.* **2017**, *49*, 1015–1024. [CrossRef] [PubMed]
4. Silva, L.H.P.; Paulino, P.V.R.; Assis, G.J.F.; Assis, D.E.F.; Estrada, M.M.; Silva, M.C.; Silva, J.C.; Martins, T.S.; Valadares Filho, S.C.; Paulino, M.F. Effect of post-weaning growth rate on carcass traits and meat quality of Nellore cattle. *Meat Sci.* **2017**, *123*, 192–197. [CrossRef]
5. Dong, G.-F.; Zou, Q.; Wang, H.; Huang, F.; Liu, X.-C.; Chen, L.; Yang, C.-Y.; Yang, Y.-O. Conjugated linoleic acid differentially modulates growth, tissue lipid deposition, and gene expression involved in the lipid metabolism of grass carp. *Aquaculture* **2014**, *432*, 181–191. [CrossRef]
6. Schoonmaker, J.P.; Cecava, M.J.; Faulkner, D.B.; Fluharty, F.L.; Zerby, H.N.; Loerch, S.C. Effect of source of energy and rate of growth on performance, carcass characteristics, ruminal fermentation, and serum glucose and insulin of early-weaned steers. *J. Anim. Sci.* **2003**, *81*, 843–855. [CrossRef]
7. Smith, S.; Blackmon, T.; Sawyer, J.; Miller, R.; Baber, J.; Morrill, J.; Cabral, A.; Wickersham, T. Glucose and acetate metabolism in bovine intramuscular and subcutaneous adipose tissues from steers infused with glucose, propionate, or acetate. *J. Anim. Sci.* **2018**, *96*, 921–929. [CrossRef]
8. Hoffmann, A.; Berça, A.S.; Cardoso, A.d.S.; Fonseca, N.V.B.; Silva, M.L.C.; Leite, R.G.; Ruggieri, A.C.; Reis, R.A. Does the effect of replacing cottonseed meal with dried distiller's grains on Nellore bulls finishing phase vary between pasture and feedlot? *Animals* **2021**, *11*, 85. [CrossRef]
9. Torrecilhas, J.A.; San Vito, E.; Fiorentini, G.; de Souza Castagnino, P.; Simioni, T.A.; Lage, J.F.; Baldi, F.; Duarte, J.M.; da Silva, L.G.; Reis, R.A. Effects of supplementation strategies during the growing phase on meat quality of beef cattle finished in different systems. *Livest. Sci.* **2021**, *247*, 104465. [CrossRef]
10. Chail, A.; Legako, J.F.; Pitcher, L.R.; Griggs, T.C.; Ward, R.E.; Martini, S.; MacAdam, J.W. Legume finishing provides beef with positive human dietary fatty acid ratios and consumer preference comparable with grain-finished beef. *J. Anim. Sci.* **2016**, *94*, 2184–2197. [CrossRef]
11. Morgan, S.A.; Huws, S.A.; Lister, S.J.; Sanderson, R.; Scollan, N.D. Phenotypic variation and relationships between fatty acid concentrations and feed value of perennial ryegrass genotypes from a breeding population. *Agronomy* **2020**, *10*, 343. [CrossRef]
12. Detmann, E.; Paulino, M.F.; Valadares Filho, S.C.; Huhtanen, P. Nutritional aspects applied to grazing cattle in the tropics: A review based on Brazilian results. *Semin. Ciênc. Agrár.* **2014**, *35*, 2829–2854. [CrossRef]
13. Valadares Filho, S.C.; Marcondes, M.I.; Chizzotti, M.; Paulino, P. *Nutrient Requirements of Zebu and Crossbred Cattle BR-CORTE*, 3rd ed.; UFV Suprema Gráfica Ltda: Viçosa, Brazil, 2016; pp. 1–327.
14. Bligh, E.G.; Dyer, W.J. A rapid method of total lipid extraction and purification. *Can. J. Biochem. Physiol.* **1959**, *37*, 911–917. [CrossRef]
15. Christie, W. Extraction and hydrolysis of lipids and some reactions of their fatty acid components. *Handb. Chromatogr. Lipids* **1984**, *1*, 33–46.
16. Martin, D.B.; Horning, M.G.; Vagelos, P.R. Fatty acid synthesis in adipose tissue: I. Purification and properties of a long chain fatty acid-synthesizing system. *J. Biol. Chem.* **1961**, *236*, 663–668. [CrossRef] [PubMed]
17. Bergmeyer, H.U.; Bernt, E. Malate dehydrogenase and isocitrate dehydrogenase. In *Bergmeyer HU*; Bergmeyer, H.U., Ed.; Methods of enzymatic analysis; Academic Press: New York, NY, USA, 1974; pp. 613–631.
18. Bergmeyer, H.U.; Bernt, E. D-Glucose determination with hexokinase and glucose-6 phosphate dehydrogenase. In *Bergmeyer HU*; Bergmeyer, H.U., Ed.; Academic Press: New York, NY, USA, 1974; pp. 1196–1201+1974.
19. Vandesompele, J.; De Preter, K.; Pattyn, F.; Poppe, B.; Van Roy, N.; De Paepe, A.; Speleman, F. Accurate normalization of real-time quantitative RT-PCR data by geometric averaging of multiple internal control genes. *Genome Biol.* **2002**, *3*, research0034.1. [CrossRef]
20. Pfaffl, M.W. A new mathematical model for relative quantification in real-time RT-PCR. *Nucl. Acids Res.* **2001**, *29*, e45. [CrossRef]
21. Bindea, G.; Mlecnik, B.; Hackl, H.; Charoentong, P.; Tosolini, M.; Kirilovsky, A.; Fridman, W.-H.; Pagès, F.; Trajanoski, Z.; Galon, J. ClueGO: A Cytoscape plug-in to decipher functionally grouped gene ontology and pathway annotation networks. *Bioinformatics* **2009**, *25*, 1091–1093. [CrossRef]
22. Lancaster, P.A.; Krehbiel, C.R.; Horn, G.W. Ameta-analysis of effects of nutrition and management during the stocker and backgrounding phase on subsequent finishing performance and carcass characteristics. *Prof. Anim. Sci.* **2014**, *30*, 602–612. [CrossRef]
23. Segabinazzi, L.R.; Menezes, L.F.G.d.; Silva, C.E.K.d.; Clederson Martinello, B.B.; Molinete, M.L. Diurnal ingestive behavior of Holstein calves reared in different systems: Feedlot or pasture. *Acta Scientiarum. Anim. Sci.* **2014**, *36*, 225–231. [CrossRef]

24. Marcondes, M.I.; da Silva, A.L.; Gionbelli, M.P.; Valadares Filho, S.C.; Costa e Silva, L.F.; Rotta, P.P.; Chizzotti, M.L.; Prados, L.F. Energy requirements for beef cattle. In *Nutrient Requirements of Zebu and Crossbred Cattle*, 3rd ed.; Valadares Filho, S.C., Costa e Silva, L.F., Gionbelli, M.P., Rotta, P.P., Marcondes, M.I., Chizzotti, M.L., Prados, L.F., Eds.; Grafica Suprema: Viçosa, Brazil, 2016; pp. 157–184.
25. Laliotis, G.P.; Bizelis, I.; Rogdakis, E. Comparative approach of the de novo fatty acid synthesis (lipogenesis) between ruminant and non ruminant mammalian species: From biochemical level to the main regulatory lipogenic genes. *Curr. Genom.* **2010**, *11*, 168–183. [CrossRef] [PubMed]
26. Ingle, D.L.; Bauman, D.E.; Garrigus, U.S. Lipogenesis in the ruminant: In vitro study of tissue sites, carbon source and reducing equivalent generation for fatty acid synthesis. *J. Nutr.* **1972**, *102*, 609–616. [CrossRef]
27. Lee, J.-H.; Go, Y.; Kim, D.-Y.; Lee, S.H.; Kim, O.-H.; Jeon, Y.H.; Kwon, T.K.; Bae, J.-H.; Song, D.-K.; Rhyu, I.J. Isocitrate dehydrogenase 2 protects mice from high-fat diet-induced metabolic stress by limiting oxidative damage to the mitochondria from brown adipose tissue. *Exp. Mol. Med.* **2020**, *52*, 238–252. [CrossRef] [PubMed]
28. Wood, J.D.; Enser, M.; Fisher, A.V.; Nute, G.R.; Sheard, P.R.; Richardson, R.I.; Hughes, S.I.; Whittington, F.M. Fat deposition, fatty acid composition and meat quality: A review. *Meat Sci.* **2008**, *78*, 343–358. [CrossRef]
29. Smith, S.B.; Gill, C.A.; Lunt, D.K.; Brooks, M.A. Regulation of fat and fatty acid composition in beef cattle. *Asian-Australas. J. Anim. Sci.* **2009**, *22*, 1225–1233. [CrossRef]
30. Frank, D.; Ball, A.; Hughes, J.; Krishnamurthy, R.; Piyasiri, U.; Stark, J.; Watkins, P.; Warner, R. Sensory and flavor chemistry characteristics of Australian beef: Influence of intramuscular fat, feed, and breed. *J. Agric. Food Chem.* **2016**, *64*, 4299–4311. [CrossRef] [PubMed]
31. Smith, S.B.; Lunt, D.K.; Chung, K.Y.; Choi, C.B.; Tume, R.K.; Zembayashi, M. Adiposity, fatty acid composition, and delta-9 desaturase activity during growth in beef cattle. *Anim. Sci. J.* **2006**, *77*, 478–486. [CrossRef]
32. Fruet, A.P.B.; Stefanello, F.S.; Júnior, A.G.R.; de Souza, A.N.M.; Tonetto, C.J.; Nörnberg, J.L. Whole grains in the finishing of culled ewes in pasture or feedlot: Performance, carcass characteristics and meat quality. *Meat Sci.* **2016**, *113*, 97–103. [CrossRef]
33. Abdelhadi, L.O.; Santini, F.J.; Gagliostro, G.A. Corn silage or high moisture corn supplements for beef heifers grazing temperate pastures: Effects on performance, ruminal fermentation and in situ pasture digestion. *Anim. Feed Sci. Technol.* **2005**, *118*, 63–78. [CrossRef]
34. Wood, J.; Enser, M. Manipulating the fatty acid composition of meat to improve nutritional value and meat quality. In *New Aspects of Meat Quality*, 8th ed.; Elsevier: Cambridge, UK, 2017; pp. 501–535.
35. Noci, F.; O'kiely, P.; Monahan, F.J.; Stanton, C.; Moloney, A.P. Conjugated linoleic acid concentration in M. Longissimus dorsi from heifers offered sunflower oil-based concentrates and conserved forages. *Meat Sci.* **2005**, *69*, 509–518. [CrossRef]
36. Krusinski, L.; Maciel, I.C.F.; Sergin, S.; Goeden, T.; Schweihofer, J.P.; Singh, S.; Rowntree, J.E.; Fenton, J.I. Fatty acid and micronutrient profile of longissimus lumborum from Red Angus and Red Angus x Akaushi cattle finished on grass or grain. *Foods* **2022**, *11*, 3451. [CrossRef] [PubMed]
37. Silva, R.R.; do Prado, I.N.; da Silva, F.F.; Rotta, P.P.; Rodrigues, L.B.O.; do Prado, R.M.; de Carvalho Mesquita, B.M.A.; Alba, H.D.R.; de Carvalho, G.G.P. Fatty acid profile and chemical composition of meat from Nellore steers finished on pasture with different amounts of supplementation. *Can. J. Anim. Sci.* **2021**, *101*, 558–566. [CrossRef]
38. Teixeira, P.D.; Oliveira, D.M.; Chizzotti, M.L.; Chalfun-Junior, A.; Coelho, T.C.; Gionbelli, M.; Paiva, L.V.; Carvalho, J.R.R.; Ladeira, M.M. Subspecies and diet affect the expression of genes involved in lipid metabolism and chemical composition of muscle in beef cattle. *Meat Sci.* **2017**, *133*, 110–118. [CrossRef] [PubMed]
39. Obsen, T.; Faergeman, N.J.; Chung, S.; Martinez, K.; Gobern, S.; Loreau, O.; Wabitsch, M.; Mandrup, S.; McIntosh, M. Trans-10, cis-12 conjugated linoleic acid decreases de novo lipid synthesis in human adipocytes. *J. Nutr. Biochem.* **2012**, *23*, 580–590. [CrossRef] [PubMed]
40. Rodríguez-Cruz, M.; Serna, D.S. Nutrigenomics of ω-3 fatty acids: Regulators of the master transcription factors. *Nutrition* **2017**, *41*, 90–96. [CrossRef]
41. Yang, C.; Ahmad, A.A.; Bao, P.J.; Guo, X.; Wu, X.Y.; Liu, J.B.; Chu, M.; Liang, C.N.; Pei, J.; Long, R.J. Increasing dietary energy level improves growth performance and lipid metabolism through up-regulating lipogenic gene expression in yak (*Bos grunniens*). *Anim. Feed Sci. Technol.* **2020**, *263*, 114455. [CrossRef]
42. Botolin, D.; Wang, Y.; Christian, B.; Jump, D.B. Docosahexaneoic acid (22:6,n-3) regulates rat hepatocyte SREBP-1 nuclear abundance by Erk- and 26S proteasome-dependent pathways. *J. Lipid Res.* **2006**, *47*, 181–192. [CrossRef]
43. Flowers, M.T.; Ntambi, J.M. Stearoyl-CoA desaturase and its relation to high-carbohydrate diets and obesity. *Biochim. Biophys. Acta Mol. Cell Biol. Lipids* **2009**, *1791*, 85–91. [CrossRef]
44. Bionaz, M.; Chen, S.; Khan, M.J.; Loor, J.J. Functional Role of PPARs in Ruminants: Potential Targets for Fine-Tuning Metabolism during Growth and Lactation. *PPAR Res.* **2013**, *2013*, 684159. [CrossRef]
45. Cronan, J.E., Jr.; Waldrop, G.L. Multi-subunit acetyl-CoA carboxylases. *Prog. Lipid Res.* **2002**, *41*, 407–435. [CrossRef]
46. Hertzel, A.V.; Smith, L.A.; Berg, A.H.; Cline, G.W.; Shulman, G.I.; Scherer, P.E.; Bernlohr, D.A. Lipid metabolism and adipokine levels in fatty acid-binding protein null and transgenic mice. *Am. J. Physiol.-Endocrinol. Metab.* **2006**, *290*, E814–E823. [CrossRef]
47. Yang, C.; Liu, J.; Wu, X.; Bao, P.; Long, R.; Guo, X.; Ding, X.; Yan, P. The response of gene expression associated with lipid metabolism, fat deposition and fatty acid profile in the longissimus dorsi muscle of Gannan yaks to different energy levels of diets. *PLoS ONE* **2017**, *12*, e0187604. [CrossRef] [PubMed]

48. Bunger, M.; van den Bosch, H.M.; van der Meijde, J.; Kersten, S.; Hooiveld, G.J.; Muller, M. Genome-wide analysis of PPARα activation in murine small intestine. *Physiol. Genom.* **2007**, *30*, 192–204. [CrossRef] [PubMed]
49. Ladeira, M.M.; Schoonmaker, J.P.; Swanson, K.C.; Duckett, S.K.; Gionbelli, M.P.; Rodrigues, L.M.; Teixeira, P.D. Nutrigenomics of marbling and fatty acid profile in ruminant meat. *Animal* **2018**, *12*, s282–s294. [CrossRef] [PubMed]

metabolites

MDPI

Article

Early Weaning Possibly Increases the Activity of Lipogenic and Adipogenic Pathways in Intramuscular Adipose Tissue of Nellore Calves

Ariane Enara Pedro [1], Juliana Akamine Torrecilhas [2], Rodrigo Nazaré Santos Torres [2], Germán Darío Ramírez-Zamudio [3], Welder Angelo Baldassini [2], Luis Artur Loyola Chardulo [2], Rogério Abdallah Curi [2], Gustavo Henrique Russo [1], Juliane Arielly Napolitano [4], Gustavo Lucas Bezerra Tinoco [1], Thiago Barcaça Mariano [4], Jordana Luiza Caixeta [4], Philipe Moriel [5] and Guilherme Luis Pereira [1,2,*]

1 College of Agronomics and Veterinary Sciences, University of São Paulo State Júlio de Mesquita Filho, Jaboticabal 14884-900, Brazil; ariane.enara@unesp.br (A.E.P.); gustavo.russo@unesp.br (G.H.R.); gustavo.tinoco@unesp.br (G.L.B.T.)
2 College of Veterinary and nimal Science, University of São Paulo State Júlio de Mesquita Filho, Botucatu 18618-687, Brazil; juliana.akamine@unesp.br (J.A.T.); rodrigo.chaves@unesp.br (R.N.S.T.); w.baldassini@unesp.br (W.A.B.); luis.artur@unesp.br (L.A.L.C.); rogerio.curi@unesp.br (R.A.C.)
3 College of Animal Science and Foods Engineering, University of São Paulo, Pirassununga 13635-900, Brazil; germanramvz@gmail.com
4 College of Agronomic Science, University of São Paulo State Júlio de Mesquita Filho, Botucatu 18610-034, Brazil; juliane.arielly@unesp.br (J.A.N.); thiago.barcaca@unesp.br (T.B.M.); jordana.caixeta@unesp.br (J.L.C.)
5 Institute of Food and Agricultural Sciences, University of Florida, Gainesville, FL 32603, USA; pmoriel@ufl.edu
* Correspondence: guilherme.luis@unesp.br

Abstract: This study aimed to evaluate by wide-expression profile analysis how early weaning at 120 days can alter the skeletal muscle metabolism of calves supplemented with a concentrated diet until the growth phase. *Longissimus thoracis* muscle samples were obtained by biopsy from two groups of calves, early weaned (EW; n = 8) and conventionally weaned (CW; n = 8) at two different times (120 days of age—T1 [EW] and 205 days of age—T2 [CW]). Next, differential gene expression analysis and functional enrichment of metabolic pathways and biological processes were performed. The results showed respectively 658 and 165 differentially expressed genes when T1 and T2 were contrasted in the early weaning group and when early and conventionally weaned groups were compared at T2. The *FABP4*, *SCD1*, *FASN*, *LDLR*, *ADIPOQ*, *ACACA*, *PPARD*, and *ACOX3* genes were prospected in both comparisons described above. Given the key role of these differentially expressed genes in lipid and fatty acid metabolism, the results demonstrate the effect of diet on the modulation of energy metabolism, particularly favoring postnatal adipogenesis and lipogenesis, as well as a consequent trend in obtaining better quality cuts, as long as an environment for the maintenance of these alterations until adulthood is provided.

Keywords: gene expression; functional enrichment; RNA-Seq

1. Introduction

Zebu breeds, including Nellore cattle, are tolerant to adverse environmental conditions such as heat, humidity, and endo- and ectoparasites typical of tropical climates, maintaining reasonable production rates [1]. However, the productive and meat quality indices of these breeds are lower compared to taurine cattle [2], these traits being complex and of low heritability because environmental factors mainly influence them or because they are difficult to measure [3,4]. Although it is a challenge, zebu breeders around the world have been using improved tools such as the use of massive information from molecular markers for genetic selection and reproductive biotechnologies in an attempt to improve production rates and the value of products in the international market [2].

Other strategies that directly influence the performance of animals correspond to nutritional management in the different stages of growth and include offering concentrate via creep-feeding [5] to providing different sources and levels of supplementation during the rearing phase [6]. Creep-feeding as a nutritional strategy not only improves weight gain and carcass quality by slaughtering animals earlier [5] but can also have indirect positive effects on dam performance, as it can reduce the production of milk due to a lower sucking stimulation of the calves caused by the replacement with concentrate [7,8]. In addition to the use of creep-feeding, early weaning as a management strategy for cows, mainly primiparous, aims to reduce the negative impact of lactation on nutritional status, allowing recovery of body condition for the next breeding season [9–11].

Traditionally, Nellore cows, the basis of Brazil's breeding herd, are weaned approximately at the seventh month of lactation [12]. However, milk production after the third month of lactation does not meet the nutritional requirements to express the genetic potential of the offspring [13]. As for calves, early weaning per se has shown negative responses on post-weaning growth [14,15]. Therefore, nutrition after this type of management has to be adequate not to compromise the calves' growth [11,16,17]. Studies indicate that early weaning associated with adequate nutritional management can lead to accelerated animal growth in the subsequent production [18–20], changes in skeletal muscle energy metabolism [21], and cellular dynamics in adipose tissue [22–24], thus improving meat quality. Although some results are still conflicting, the magnitude of metabolic modulation is directly related to the level and balance of nutrients in diets offered to calves [25]. However, research to understand skeletal muscle tissue metabolism affected by management strategies such as early weaning and nutritional plans applied during the growth phase of calves is limited. Therefore, zebu cattle need to be improved.

It is hypothesized that early weaning attended by moderate supplementation with concentrate until the traditional weaning time for beef cattle does not affect the growth of calves. However, it alters the gene regulation that regulates metabolic pathways related to fat deposition in skeletal muscle tissue, improving meat quality. Therefore, this study aimed to evaluate the effect of early weaning associated with a post-weaning supplementation protocol and compare it with traditional weaning on growth, changes in biological processes, and metabolic pathways regulated by gene expression profiles in skeletal muscle tissue.

2. Materials and Methods

2.1. Samples and Weight Recording

Forty male Nellore calves born in the same month to contemporary cows of the same calving order were used. A field experiment was carried out on a duly regularized commercial farm, located in the State of Mato Grosso (Pantanal biome), Brazil. The animals were submitted to two different treatments, in which 20 male calves were kept with their respective dams in the same paddock on *Brachiaria* spp. pasture from birth until 205 days of age (~30 weeks) (Time point 2—T2 or Moment 2—M2) when they were conventionally weaned (CW), and 20 males calves were weaned early (EW) at 120 days of age (~16 weeks) (Time point 1—T1 or moment—M2), reassigned to the same paddock with Tifton 85 pasture and supplemented with concentrate (Figure 1) until 205 days of age (~30 weeks).

All calves were weighed after a 6 h fast at the beginning of the early weaning phase at 120 days (initial body weight, BWi) and at conventional weaning, at 205 days of age (weaning weight, WW). Average daily weight gain (ADG) was then calculated based on BWi and WW for both groups.

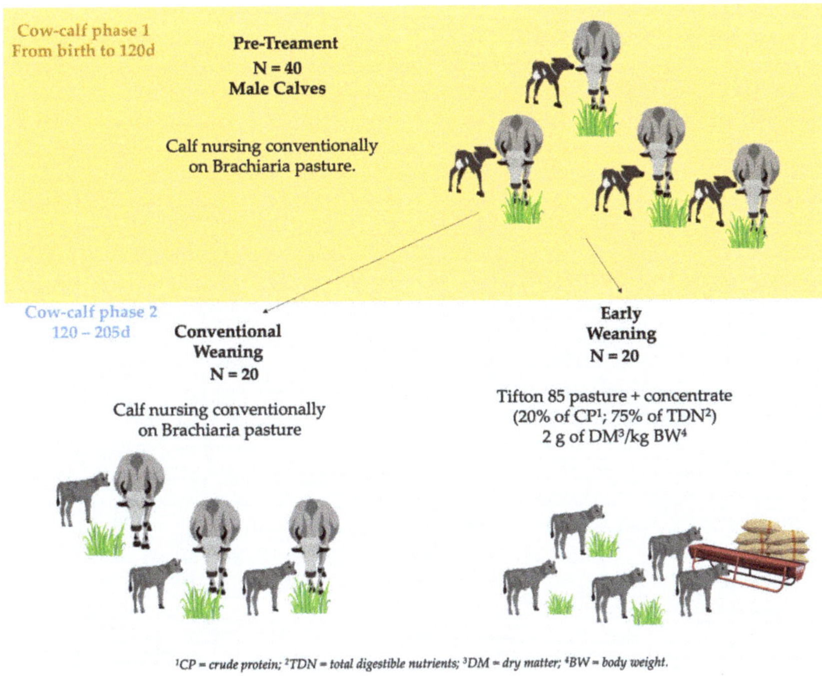

Figure 1. Supplementation and diets consisting of natural forage, soybean meal, corn, additives, and minerals are offered to Nellore cattle at different developmental stages and subjected to different weaning protocols.

2.2. Growing Performance Statistics

To detect significant differences in weights between treatments and time points, fulfilling all assumptions, analysis of variance (ANOVA) was conducted using a linear model with random residuals, in which the fixed effects of treatment (EW and CW), the time point of weaning (T1 and T2), and their interaction on the response variable (WW) were tested. Significant differences between means of specific groups were evaluated using the Tukey test for multiple comparisons at a level of significance of 5%. The difference in ADG between treatments (EW and CW) was obtained using an independent t-test after observation of a normal distribution of the ADG records and homogeneity of variances between the two groups.

2.3. Collection of Muscle Tissue and Total RNA Extraction

About 1 g samples of the *Longissimus thoracis* (LT) muscle were collected by biopsy between the twelfth and thirteenth rib during early and conventional weaning. For this purpose, a local anesthetic was administered subcutaneously, and the site was shaved and cleaned. Next, a 1 cm incision was made with a scalpel blade, and a sterile biopsy needle was used to obtain 1 g of muscle tissue. The sample was immediately transferred to liquid nitrogen and stored in an ultra-freezer at −80 °C.

Total RNA was extracted individually from 100 mg of LT muscle using TRIzol® (Life Technologies, Carlsbad, CA, USA) in accordance with the manufacturer's instructions, and its quality was analyzed in a Bioanalyzer 2100® (Agilent, Santa Clara, CA, USA). To ensure adequate total RNA quality, only samples with an RNA integrity number (RIN) ≥ 7 were used for sequencing.

2.4. RNA Sequencing

After total RNA extraction, the isolated mRNA was submitted to purification and fragmentation using oligo-dT beads for the construction of 32 cDNA libraries. The libraries consisted of eight samples of each treatment (CW and EW) collected at T1 and T2, respectively. The cDNA libraries for each sample were prepared and multiplexed from 2 μg total RNA using the Illumina Stranded mRNA Prep kit (Illumina, San Diego, CA, USA) according to the Illumina Stranded mRNA Prep, Ligation Reference Guide (Illumina, USA). The average size of the libraries was determined by quantitative PCR (RT-qPCR) using the KAPA Library Quantification kit (KAPA Biosystems, Wilmington, MA, USA) in a Bioanalyzer 2100® (Agilent, Santa Clara, CA, USA). Clustering and sequencing were performed with the NextSeq 550® System (Illumina, USA) in order to produce 150 bp paired-end (PE) reads, with an expected average coverage of 10 million PE reads per sample.

2.5. Mapping of Sequences to the Reference Genome

The pipeline indicating the steps and programs used to obtain the count matrix from transcriptome data is illustrated in Figure S1. First, FastQC v. 0.11.9 [26] was used to analyze the quality of raw reads. Sequencing adapters and low-quality sequences were removed using the fastp v.0.20.0 program [27], ignoring reads and quality scores of poly-G segments, characteristic of null reads. After this step, the quality of the reads was reassessed by joint graphic visualization of all FastQC outputs using the MultiQC v.1.13 program [28]. The reads were then mapped to the bovine reference genome (Bos taurus—ARS-UCD 1.2), available at http://www.ensembl.org/Bos_taurus/Info/Index/ (accessed on 24 September 2022), using STAR v.2.7.20 [29]. The reads were mapped independently for each sample and only sequences that mapped uniquely to the genome and to known chromosomes were used in the analysis of differential gene expression. Considering the generated sequences (aligned PE reads), the number of mapped sequences per gene (exons) was determined to generate a count matrix (genes × samples) using the feature counts v.2.0.3 software [30].

2.6. Identification of Differentially Expressed Genes

The edgeR v.3.38.1 package [31] of R v.4.2.1 [32] was used to identify differentially expressed genes (DEGs). First, for quality control and consistency analysis of the count data, gene expression levels were obtained by adjusting the counts to the size of the library of each sample based on counts per million (CPM) and then normalized by log2 transformation [log2(CPM)]. Genes with low CPM were removed according to the default option of the edgeR v.3.38.1 package [31] and principal component analysis (PCA) was performed using the factoextra v.1.0.7 package [33] of R v.4.2.1 [32]. To estimate the effects of the treatments on gene expression, normalization factors (size factors) were calculated for each sample using the trimmed mean of M-values (TMM) method. The negative binomial distribution hyperparameter was estimated using the robust empirical Bayes algorithm. A generalized linear model was then adjusted assuming a negative binomial distribution and quasi-likelihood was used for coefficients inference. The significance of the effect of treatment on gene expression was obtained as the likelihood ratio between the reduced and complete models for each gene, assuming that the values of each test belong to a $\chi 2$ distribution. Next, the *p*-values were adjusted by the Benjamini–Hochberg method to correct the false-discovery rate (FDR) for multiple tests. Finally, the relative expression of each gene was obtained based on the logarithmic function of the ratio between the CPM of the different contrasts, as follows: $\log 2\left(\frac{\text{mean(CPM\_A)}}{\text{mean(CPM\_B)}}\right) = \log 2(\text{FC})$, where FC is the fold change between contrasts. A log2(FC) value ≤ 0.5 or ≥ 0.5, and an adjusted *p*-value < 0.05 were defined as the thresholds for the detection of DEGs between the different contrasts (EW vs. CW at T1; EW vs. CW at T2; T1 vs. T2 in EW; T1 vs. T2 in CW). The terms "upregulated" and "downregulated" refer to EW compared to CW and to T2 compared to T1.

2.7. Enrichment of Functional Terms

Enrichment of biological processes (BP) belonging to gene ontology (GO) terms and metabolic pathways annotated in the Kyoto Encyclopedia of Genes and Genomes (KEGG) was performed using the clusterProfiler v.4.4.4 [34], enrichplot v.1.18.3 [35], and enrichR v.3.1 [36] packages of R v.4.2.1 [32]. Over-representation analysis of BP and KEGG was performed using a hypergeometric test, considering an alpha level of significance of 5% corrected by FDR for GO terms tests. The abundance of genes related to a GO term or KEGG pathway present in the DEG list was compared to that of non-DEGs.

3. Results

3.1. Pre- and Post-Early Weaning Performance

No significant differences ($p < 0.05$) in BWi, WW, or ADG estimates were observed between the EW and CW groups (Table 1). A greater variation (SD) and range (minimum and maximum) of final body weight (BWf) was observed in the EW group compared to the CW group. In general, BWf or ADG was not affected by the stress associated with early weaning. However, environmental and intrinsic factors may have caused greater than expected heterogeneity in the performance of EW calves.

Table 1. Live weight at 120 (time 1) and 205 (time 2) days of age and average daily weight gain (ADG) of early (EW) and conventionally weaned (CW) Nellore calves.

Categories	Group	Days	n	Min.	Max.	Median	Mean [1]	SD [2]
Live weight (kg)	CW	120	20	103	155	122	122.50 ± 2.961^{b}	13.241
	EW	120	20	96	141	125	123.15 ± 2.606^{b}	11.654
	CW	205	20	169	216	198	194.39 ± 3.023^{a}	12.825
	EW	205	20	156	221	191	190.32 ± 4.375^{a}	19.070
ADG (kg)	CW	205–120	20	0.553	1.047	0.853	0.845 ± 0.028^{a}	0.127
	EW	205–120	20	0.588	1.024	0.788	0.786 ± 0.030^{a}	0.133

[1] Mean = mean ± standard error; [2] SD = standard deviation. Different letters indicate significant differences (p-value > 0.05) between means by Tukey's multiple test and t for the Live Weight and ADG categories, respectively.

3.2. RNA Extraction and Sequencing

The mean total RNA concentration obtained for the 32 samples was 260.27 ng/µL and the mean RIN was 7.60 (range: 7.02 to 8.64). Among the 750 million reads obtained by sequencing (~23.60 million reads/sample), 94.3% passed the quality filters, providing about 710 million reads for alignment (~22.60 million reads/sample); of these, 94.31% had quality $\geq$ Q30 (~2.1 Gb/sample) (Table S1).

Sequencing of mRNA resulted in 321.56 million PE reads (2 × 100 bp) (~10.05 million PE reads/sample); of these, 316.5 million were uniquely mapped PE reads (Table S2). On average, 9.9 million uniquely mapped PE reads per sample were obtained, corresponding to 88.81% of all PE reads generated. Among all uniquely mapped PE reads, about 70.35% were assigned to exons (~7.9 million PE reads/sample) (Table S2) and were used for the subsequent descriptive and inferential analyses.

3.3. Count Matrix and Quality

Among 27,607 bovine genes, 16,034 were mapped to at least one PE read in one sample. After the deletion of features with low counts of PE reads already assigned using the edgeR package algorithm, 14,073 genes remained with sufficient counts within the groups, which were submitted to the analysis of differential gene expression. Figure S3 shows the size of the mapped libraries and the boxplots of the normalized read counts [log2(CPM)], with the consistent quartile distribution of samples between groups indicating good quality of the sequencing data.

A PCA analysis using measurements of the expression of each gene, reported as standardized log2CPM (Figure S2), showed that the similarity of individuals within groups was not much greater than between the different groups, indicating that the overall gene expression profile was different between the two time points and similar between individuals at each time point; however, the same is not observed when the treatments within time points are compared, particularly for the contrast of EW vs. CW at T1, which would be expected considering that the EW group had not yet been weaned. Although the division between treatments was more consistent at T2 than at T1, a better separation between individuals of the EW and CW groups at T2 would be expected. This was not observed, probably because they share expression profiles of many more similar growth-related genes than those expected to be altered by the applied treatments. Together, the first two components captured approximately 20% of the total variation observed between expression levels in the total sample (Figure S2). Figure S3 shows the results of cluster analysis illustrated as a heatmap, in which individuals (n) are grouped based on shared expression levels of the 100 most DEGs (m). These genes, in turn, are grouped according to the expression profile shared between samples. In this case, it is possible to observe the expression level of each gene in each sample and at each time point. The results show that, although the samples were not grouped in a more defined manner based on their treatments, there is a clear division between the EW and CW groups when we consider the genes most likely to be affected by the treatments.

3.4. Differential Gene Expression

Analysis of differential gene expression identified respectively 23, 165, 901, and 658 DEGs for the following contrasts: EW vs. CW at T1, EW vs. CW at T2, T1 vs. T2 in CW, and T1 vs. T2 in EW (Figure 2; Table S3). As expected, there were few prospected DEGs within T1, in contrast to the other comparisons, including that between groups in T2. Within this context, in addition to the larger number of DEGs prospected at T2 (EW vs. CW), there are also genes that play key roles in energy metabolism, such as those involved in carbohydrate and lipid metabolism, and even some that may be important for the differentiation of muscle fibers. These genes include *FABP4*, *SCD1*, *FASN*, *LDLR*, *ADIPOQ*, *ACACA*, *PPARD*, and *ACOX3*, which are classically related to energy metabolism and are modulated by lipogenic and adipogenic activity.

When differences over time (T1 vs. T2) were analyzed within each treatment (EW or CW), we found a large number of DEGs compared to those identified based on the contrasts within T1 and T2 (Figure 2). Although both treatments returned abundant findings, in addition to the DEGs commonly modulated during early mammalian development, there was broad modulation of gene expression in the EW group, probably due to the intake of the concentrated diet. Many of these genes were also observed in the contrast of EW vs. CW at T2. Logically, these DEGs were not found in CW calves between T1 and T2, which consumed the same habitual diet that may have led to the modulation of genes commonly involved in growth and genes involved in environmental responses and in the gradual change to a predominantly grazing diet.

3.5. Functional Enrichment Analysis of Differentially Expressed Genes

The list of DEGs obtained for each contrast was used for functional analysis (Figure 2). Considering the contrasts of interest (EW vs. CW at T2 and T1 vs. T2 in EW), broad activity of BP (Figure 3) and metabolic pathways (Figure 4) related to metabolism, signaling, and metabolic processes of fatty acids and lipids was observed. Among BP terms, we observed an abundance of metabolic and biosynthetic processes of lipids, fatty acids, organic acids, and mono- and carboxylic acids. In EW vs. CW at T2, sterol and cholesterol homeostasis terms were exclusively enriched based only on genes downregulated in EW (Figure 2B). Similarly, in the contrast of T1 vs. T2 in EW, the proportion of downregulated genes was considerably higher for fatty acid polysaccharides bioprocesses, fatty acid oxidation, fatty-acid beta-oxidation, and fatty acid catabolic process terms (Figure 3D).

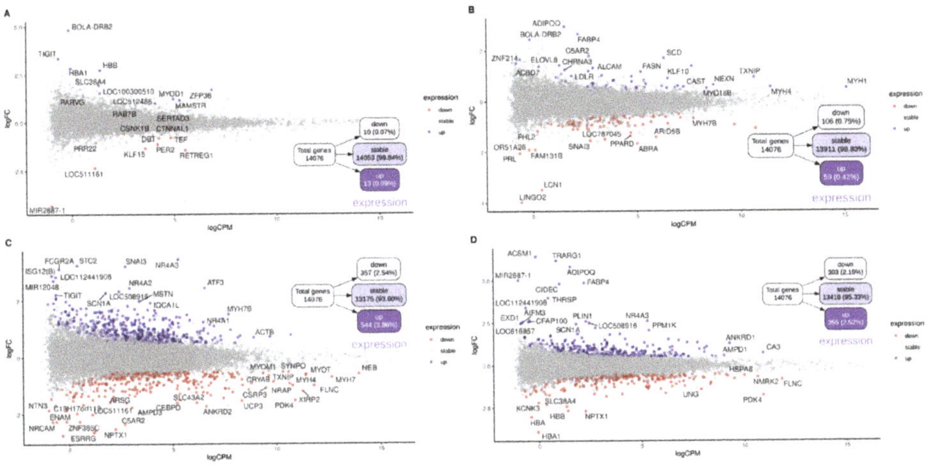

Figure 2. MD plot of different contrasts between the early weaning (EW), conventional weaning (CW), time 1 (T1), and time 2 (T2) groups. MDplot containing general expression normalised (logCPM) (x-axis), Fold Change (logCPM) (y-axis) and significative up- and down regulated differentially expressed genes(blue or red dots, respectivelly). The total number and percentage of genes and DEGs prospected are represented into boxes in treeplot for each contrasts: early weaning (EW) vs. conventional weaning (CW) at time 1 (T1) (**A**) and time 2 (T2) (**B**), and T1 vs. T2 within EW (**C**), and CW (**D**).

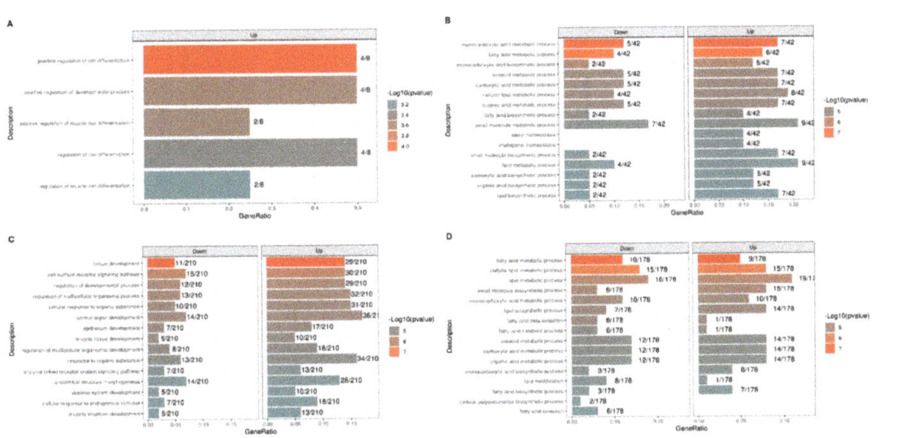

Figure 3. Biological process terms (BP) enriched from combined up- and downregulated DEGs in each contrast test: early weaning (EW) vs. conventional weaning (CW) at time 1 (T1) (**A**) and time 2 (T2) (**B**), and T1 vs. T2 within EW (**C**), and CW (**D**).

Among the enriched metabolic pathways, we highlight fatty acid metabolism, PPAR, AMPK, adipocytokine signaling pathways, fatty acid biosynthesis, and fatty acid degradation (Figure 4B,D). The last was observed exclusively in the contrast of T1 vs. T2 in EW, with only downregulated genes after EW (2). The downregulated genes include *ACOT3*, *ACOX3*, *PPARD*, *CPT1A*, *CEPT1B*, *CPT2*, *MLYCD*, and *FABP3* (Figure 5B,D; Tables S4 and S5), which are closely linked to the utilization of fat stores as energy and to antagonistic activities to lipogenesis. On the other hand, genes that play a key role in adipogenic and lipogenic metabolism were upregulated in pathways that participate in adipocyte differentiation and maturation, as well as in lipogenic activities, in addition to possible changes in myocyte energy profiles. We highlight the *FABP4*, *SCD1*, *FASN*, *PLIN1*, *LDLR*, and *ADIPOQ* genes shared between EW vs. CW at T2 and T1 vs. T2 in EW (Figures 5B,D and 6; Table S6).

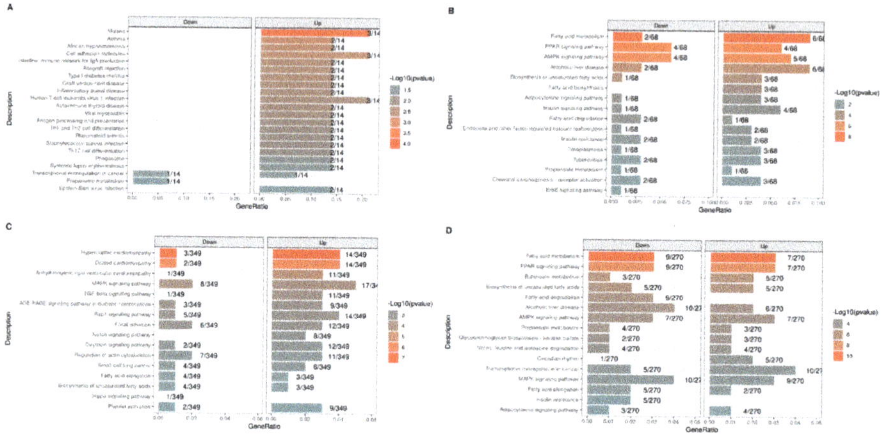

Figure 4. KEGG pathways enriched from combined up- and downregulated DEGs in each contrast test: early weaning (EW) vs. conventional weaning (CW) at time 1 (T1) (**A**) and time 2 (T2) (**B**), and T1 vs. T2 within EW (**C**) and CW (**D**).

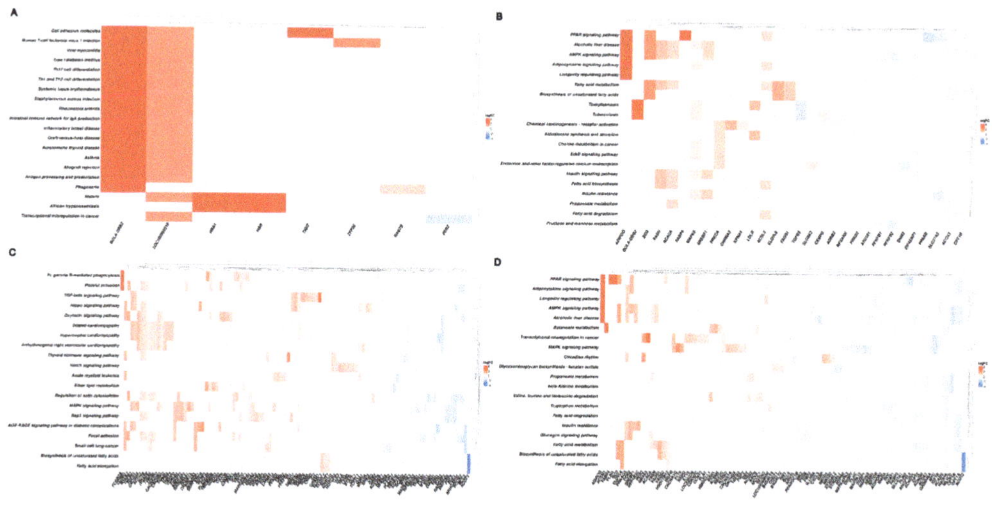

Figure 5. Clustergram of enriched KEGG pathways based on log2 (fold change) of up- and downregulated DEGs of different contrasts: early weaning (EW) vs. conventional weaning (CW) at time 1 (T1) (**A**) and time 2 (T2) (**B**), and T1 vs. T2 within EW (**C**) and CW (**D**).

A

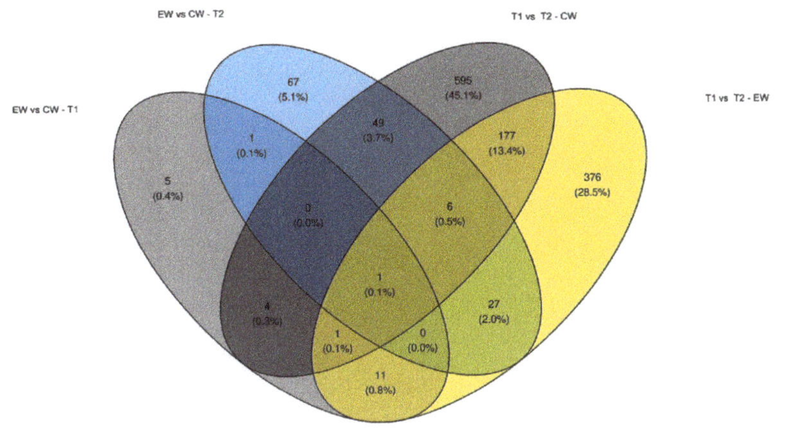

Shared DEG's between EW vs CW - T2 and T1 vs T2 - EW: SCD; FADS1; ACACA; FCGRT; SREBF1; NR1D1; CORO1C; LSS; CPT1B; CFAP70; KCNA5; MIR133A-2; KLF13; CSRP2; DYNLL1; MAMSTR; CHRNA3; FASN; ACOX3; SNAI3; MYH7B; ZNF385B; KCNQ4; CRELD2; ADIPOQ; DYNC2I2; GRIP2; MYLK4; PEG3; PRL; HSPH1; MPDZ; ELOVL6; FABP4

B

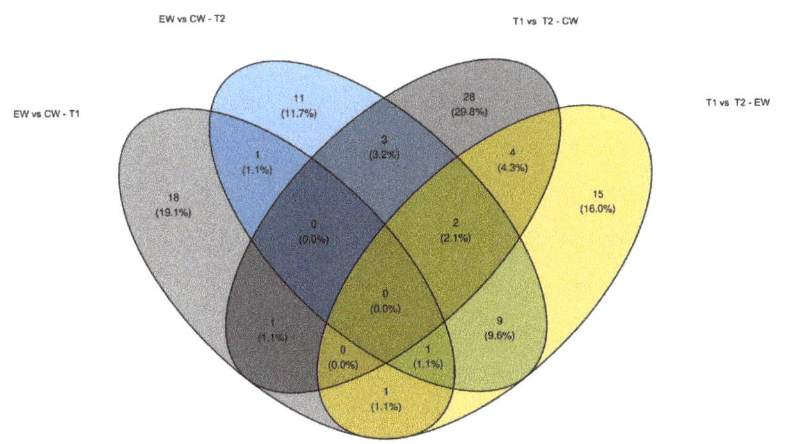

Shared KEGG pathways between EW vs CW - T2 and T1 vs T2 - EW: Fatty acid metabolism; PPAR signaling pathway; AMPK signaling pathway; Alcoholic liver disease; Biosynthesis of unsaturated fatty acids; Adipocytokine signaling pathway; Insulin signaling pathway; Fatty acid degradation; Insulin resistance; Propanoate metabolism; Longevity regulating pathway; Glucagon signaling pathway

Figure 6. Venn diagram of shared DEGs (**A**) and KEGG pathways (**B**) of all contrast tests and description of DEGs and KEGG pathways shared by EW vs. CW at T2 and T1 vs. T2 in EW. Gray colors are related to contrasts in which none of the groups were affected by the treatment. The blue color indicates contrast between EW and CW treatments at time point 2 and the yellow color indicates contrast comparing EW at time point 1 and 2.

4. Discussion

Early weaning accompanied by a high concentrate diet or supplementation shows contradictory responses regarding the performance of calves during the weaning period [37–39]. However, EW is associated with higher carcass weight, with a reduction in or lack of effect on kidney, pelvic, heart fat (KPH), and with an increase in the marbling score of the carcass [19,24,37,38,40]. Our results showed the absence of an effect of EW even when combined with supplementation on ADG compared to CW calves (205 days, no supplementation). However, the results indicate that EW, followed by supplementation, induces the maximum fat deposition potential, mainly as marbling or intramuscular fat (IMF) in breeds with a low potential for IMF deposition, as is the case of Nellore cattle.

Nellore animals are recognized for their low IMF content and tougher meat, factors that reduce their acceptance by consumers [41,42]. However, higher expression of adipogenic and lipogenic genes such as *FABP4*, *SCD1*, *FASN*, *LDLR*, *ADIPOQ*, *ACACA*, *PPARD*, and *ACOX3* was observed in the EW group. These genes are related to the enrichment of lipid and fatty acid biosynthetic pathways, while downregulated genes are associated with lipid oxidation pathways.

The deposition of IMF is a multifactorial physiological process that depends on factors related to genetics, management, and animal nutrition. The mechanisms that control adipogenesis in the fetus and postnatally in skeletal muscle in vivo remain poorly defined [43].

Although little is known about IMF deposition in ruminants, studies show that the development of marbling occurs by an increase in the number (hyperplasia) and volume of adipocytes (hypertrophy) or a combination of both processes [44], indicating that the window of opportunity for increasing intramuscular adipocytes in cattle is small. The fact that the development of marbling starts during the prenatal phase (end of gestation) and extends up to 250 postnatal days, which is related to pre-adipocyte hyperplasia [45], renders the increase in marbling in meat even more complex. Within this context, EW can be used as a strategy to accelerate the differentiation and maturation of intramuscular adipocytes. This was observed in the EW group at T2 in which the *ADIPOQ* and *THRSP* gene upregulated. The *ADIPOQ* gene is secreted by mature adipocytes and promotes adipocyte differentiation [46], and when upregulated, and associated with increased lipid deposition in mature adipocytes [47]. Meanwhile, higher expression of the THRSP gene is associated with higher marbling content in breeds such as the wagyu, which are recognized for their high marbling ability in the meat [48].

Among the metabolic pathways enriched in the EW group, the fatty acid metabolism, PPAR, AMPK, and adipocytokine signaling pathways can be highlighted. The PPAR family is associated with the regulation of adipocyte differentiation [49], with *PPARG* being one of the most studied members of this family. Its ability to induce adipocyte differentiation has been extensively investigated [50]. As suggested by [47] who evaluated EW in Angus and Angus x Simmental cattle, *PPARG* can upregulate the expression of activators of fatty acid synthesis (*THRPS*, *SREBP1*, and *INSIG1*) and fat synthesis-related enzymes (*FASN*, *SCD*, *ELOVL6*, *PCK1*, and *DGAT2*), which is consistent with our results. We also observed the upregulation of genes such as *THRPS*, *FASN*, *SCD*, and *ELOVL6* when compared to the CW group at T2 and in contrast to T2 vs. T1 in EW. These findings indicate high recruitment of adipocytes within the "marbling window" which, according to [51], occurs up to 250 days after birth. Within this context, EW acts not only by increasing the number of adipocytes in IMF but also by the deposition of fat in adipocytes of intramuscular adipose tissue, which are known to have a low rate of fatty acid synthesis and deposition [52,53]. The greater expression of the *PLIN1* gene for the EW group in addition to acting in the regulation of triglyceride synthesis, also favored the formation of large lipid droplets, by inhibiting the hydrolysis of triglyceride in adipocytes [54]. This can also be favored by the action of the PLIN1 gene, which is expressed in adipocytes. In addition to regulating the synthesis of triglycerides, this gene also participates in the formation of large lipid droplets by inhibiting the hydrolysis of triglycerides in adipocytes [54].

The decrease in the biological pathways involved in the degradation of fatty acids in response to EW cannot be solely related to the downregulation of *PPARD*. The decrease in the biological pathways involved in fatty acid degradation in response to EW might be related not only to the downregulation of *PPARD*. According to [50], in cattle, higher expression of *PPARD* acts on the stimulation of fatty acid oxidation as a mechanism to save glucose in animals receiving low-starch diets. Supplementation in EW acts as a positive signal of the animal's nutritional status, reducing fluctuations in the expression of the *THRSP* gene throughout the growth period. According to [55], the expression of this gene oscillates during the development of the animal and is sensitive to environmental and nutritional signals. The increase in *THRSP* gene expression for periods greater than 280 days of life was observed in animals with a high potential for marbling deposition [55].

In addition to the reduction in lipid oxidation in the EW group, it is possible that early grain intake by animals of this group reduces the competition for nutrients destined for fat and muscle deposition pathways in the carcass. During the growth phase, nutrients are partitioned in favor of bone and muscle growth, while the fat deposition rate is relatively low [56]. We observed upregulation of *MYOD1* in the EW group at T1 (up to 120 days). This gene acts together with *MyF5* as the first myogenic regulatory factor and is necessary for the determination of the myogenic lineage [57]. However, in mature muscle tissue, it has been suggested that the upregulation of *Myf5* [58,59] and *MYOD1* [60] is related to the proliferation and differentiation of satellite cells for hypertrophy, respectively.

In the control group, upregulation of the myostatin gene (*MSTN*) was observed for the T2 vs. T1 contrast, which may indicate muscle growth restriction in these animals. The *MSTN* gene is mainly expressed in muscle, where it acts as a negative regulator of skeletal muscle growth and development [61]. Its detection in adipose tissue is associated with the regulation of the development of this tissue [62]. In addition to inhibiting adipogenesis in pre-adipocytes [63], the high expression of *MSTN* and *PPARG* reduces IMF content and the size of adipocytes.

Our results help to understand the mechanism associated with the response to EW according to [24,37,38] EW has the potential to stimulate the deposition of IMF, with a minimal increase in subcutaneous fat or even KPH, which is recognized as a "carcass contamination fat" and is accompanied by an increase in carcass weight. The positive effect of EW on carcass weight and IMF content can be associated with the positive effect of EW on energy metabolism in the growth phase, which was accompanied by the stimulation of adipogenic and lipogenic pathways, without affecting the metabolic pathways associated with muscle growth. Our results also indicate the possibility of reducing saturated fatty acids in the meat of EW animals. Within this context, genes associated with the *de novo* biosynthesis and unsaturation of fatty acids such as *SCD*, *FASN*, *ELOVL6*, *ACACA*, and *FABP4* were upregulated in animals of the EW group. Furthermore, upregulated *FABP4*, *G6PD*, *FASN*, and *ACACA* act on the terminal differentiation of adipocytes [64,65]. *FASN* and *ACACA* together perform de novo synthesis of fatty acids [43], such as palmitate from acetyl-CoA and malonyl-CoA [66]. The palmitate (C16) is used as a substrate for elongase, an enzyme encoded by the *ELOVL6* gene that catalyzes the elongation of saturated and monounsaturated fatty acids with 12, 14, and 16 carbons, in which C18 fatty acids are the final product [46]. An increase in the energy and/or starch content of cattle diets is associated with higher expression of the *SCD* gene [67] that encodes Δ9 desaturase, an enzyme that converts saturated fatty acids into cis-9 monounsaturated fatty acids [52]. When upregulated, all of these genes are associated with greater IMF deposition in cattle, as observed for the *SCD* gene [52,68], which was also found to stimulate the expression of *FASN*, *ACACA*, and *FABP4* [69]. Higher expression of *FABP4* is not only associated with an increase in IMF content but also in triacylglycerol [70], and stimulates adipocyte differentiation [71].

Early weaning has the potential to alter the expression of genes that are important for increasing the IMF content of Nellore meat. This breed has been recognized for having high lipid turnover, as observed by [66] who compared Nellore and Angus animals receiving the same finishing diet. Within this context, EW favors a reduction in the finishing period of animals with low-fat deposition capacity by stimulating adipogenic and lipogenic pathways and increasing fat deposition, which may improve meat qualitative parameters such as tenderness and juiciness.

5. Conclusions

Early weaning associated with supplementation did not affect weight gain compared to traditionally weaned calves. However, early weaning, associated with an adequate nutritional strategy, positively affects regulating pathways related to energy metabolism, adipogenesis, and lipogenesis in skeletal muscle tissue increases PPAR, AMPK, and the adiponectin signaling pathway and biosynthesis of mono- and polyunsaturated fat acids. Therefore, this management and nutritional strategy can potentially increase intramuscular fat in Nellore beef.

Supplementary Materials: The following supporting information can be downloaded at: https://www.mdpi.com/article/10.3390/metabo13091028/s1, Supplementary Material S1: Alignment Workflow scheme and descriptive statistics and supplementary DEG results; Supplementary Material S2: Diferentially expressed genes; Supplementary Material S3: Enriched ontology genes terms to biological process; Supplementary Material S4: Enriched KEGG pathways; Supplementary Material S5: Differentially expressed genes shared between contrasts.

Author Contributions: Conceptualization, G.L.P., R.A.C. and L.A.L.C.; methodology, G.L.P., R.N.S.T. and G.D.R.-Z.; validation, J.A.T., W.A.B. and A.E.P.; formal analysis; investigation, G.L.P. and A.E.P.; resources, G.L.P.; data curation, P.M., G.H.R. and G.L.B.T.; writing—original draft preparation, A.E.P., P.M., T.B.M., J.L.C. and J.A.N.; writing—review and editing, A.E.P., P.M., G.L.P., R.N.S.T. and G.D.R.-Z.; visualization, A.E.P.; supervision, P.M., G.L.P. and J.A.T.; project administration, G.L.P. and J.A.T.; funding acquisition, G.L.P. All authors have read and agreed to the published version of the manuscript.

Funding: This research was funded by São Paulo Research Foundation (FAPESP) grant number 2019/12851-1.

Institutional Review Board Statement: The animal study protocol was approved by the Institutional Ethics Committee of the College of Veterinary and Animal Science of the University of São Paulo State, Brazil (protocol code 0190/2020 on 16 December 2020).

Informed Consent Statement: Not applicable.

Data Availability Statement: The data presented in this study are available on request from the corresponding author. The data are not publicly available due to privacy and patent intent.

Acknowledgments: The authors thank the São Paulo Research Foundation (FAPESP, grant 2019/12851-1).

Conflicts of Interest: The authors declare no conflict of interest.

References

1. Beatty, D.T.; Barnes, A.; Taylor, E.; Pethick, D.; McCarthy, M.; Maloney, S.K. Physiological responses of Bos taurus and Bos indicus cattle to prolonged, continuous heat and humidity. *J. Anim. Sci.* **2006**, *84*, 972–985. [CrossRef]
2. Burrow, H.M. Importance of adaptation and genotype × environment interactions in tropical beef breeding systems. *Animal* **2012**, *6*, 729–740. [CrossRef] [PubMed]
3. Kemper, K.E.; Bowman, P.J.; Pryce, J.E.; Hayes, B.J.; Goddard, M.E. Long-term selection strategies for complex traits using high-density genetic markers. *J. Dairy Sci.* **2012**, *95*, 4646–4656. [CrossRef]
4. Mrode, R.; Ojango, J.M.K.; Okeyo, A.M.; Mwacharo, J.M. Genomic selection and use of molecular tools in breeding programs for indigenous and crossbred cattle in developing countries: Current status and future prospects. *Front. Genet.* **2019**, *9*, 694. [CrossRef]
5. Carvalho, V.V.; Paulino, M.F.; Detmann, E.; Valadares Filho, S.C.; Lopes, S.A.; Rennó, L.N.; Sampaio, C.B.; Silva, A.G. A meta-analysis of the effects of creep-feeding supplementation on performance and nutritional characteristics by beef calves grazing on tropical pastures. *Livest. Sci.* **2019**, *227*, 175–182. [CrossRef]
6. Tambara, A.A.C.; Härter, C.J.; Rabelo, C.H.S.; Kozloski, G.V. Effects of supplementation on production of beef cattle grazing tropical pastures in Brazil during the wet and dry seasons: A meta-analysis. *Rev. Bras. De Zootec.* **2021**, *50*, e20210020. [CrossRef]

7. Fordyce, G.; Cooper, N.J.; Kendall, I.E.; O'Leary, B.M.; Faveri, J.D.E.; De Faveri, J. Creep feeding and prepartum supplementation effects on growth and fertility of Brahman-cross cattle in the dry tropics. *Aust. J. Exp. Agric.* **1996**, *36*, 389–395. [CrossRef]
8. Henriques, L.T.; Valadares Filho, S.D.C.; Fonseca, M.A.; Paulino, P.V.R.; Detmann, E.; Valadares, R.F.D. Evaluation of non-linear models and the effects of primiparous cows and calves intake on the lactation curve of Nelore cows. *Rev. Bras. De Zootec.* **2011**, *40*, 1287–1295. [CrossRef]
9. Lusby, K.S.; Wettemann, R.P.; Turman, E.J. Effects of early weaning calves from first-calf heifers on calf and heifer performance. *J. Anim. Sci.* **1981**, *53*, 1193–1197. [CrossRef]
10. Barnes, K.C.; Selk, G.E.; Lusby, K.S. Early weaning fall-born beef calves for improved rebreeding or decreased winter feed. *Okla. Agric. Exp. Sta. Res. Rep.* **1996**, *951*, 85–87.
11. Cattelam, J. Desmame precoce na podução de bovinos de corte. *Rev. De Ciências Agroveterinárias* **2014**, *13*, 190–198.
12. Ferraz, J.B.S.; de Felício, P.E. Production systems–An example from Brazil. *Meat Sci.* **2010**, *84*, 238–243. [CrossRef]
13. Costa e Silva, L.F.; Engle, T.E.; de Valadares Filho, S.; Rotta, P.P.; Villadiego, F.A.C.; Silva, F.A.S.; Martins, E.C.; Silva, L.H.R.; Paulino, M.F. Nellore cows and their calves during the lactation period: Performance, intake, milk composition, and total apparent digestibility. *Trop. Anim. Health Prod.* **2015**, *47*, 735–741. [CrossRef]
14. Wardrop, I.D. The effects of the plane of nutrition in early post-natal life on the subsequent growth and development of cattle. *Aust. J. Agric. Res.* **1966**, *17*, 375–385. [CrossRef]
15. Restle, J.; Polli, V.A.; Alves Filho, D.C.; Senna, D.B.D.; Vaz, R.Z.; Bernardes, R.A.C.; Silva, J.H.S.D. Desenvolvimento de bovinos de corte de diferentes grupos genéticos desmamados aos 3 ou 7 meses de idade. *Rev. Bras. De Zootec.* **1999**, *28*, 1023–1030. [CrossRef]
16. Arthington, J.D.; Spears, J.W.; Miller, D.C. The effect of early weaning on feedlot performance and measures of stress in beef calves. *J. Anim. Sci.* **2005**, *83*, 933–939. [CrossRef] [PubMed]
17. Vaz, R.Z.; Lobato, J.F.P. Efeito da idade do desmame no desenvolvimento de novilhas de corte até os 14/15 meses de idade. *Rev. Bras. De Zootec.* **2010**, *39*, 289–298. [CrossRef]
18. Neville, W.E., Jr.; McCormick, W.C. Performance of early-and normal-weaned beef calves and their dams. *J. Anim. Sci.* **1981**, *52*, 715–724. [CrossRef]
19. Fluharty, F.L.; Loerch, S.C.; Turner, T.B.; Moeller, S.J.; Lowe, G.D. Effects of weaning age and diet on growth and carcass characteristics in steers. *J. Anim. Sci.* **2000**, *78*, 1759–1767. [CrossRef] [PubMed]
20. Myers, S.E.; Faulkner, D.B.; Ireland, F.A.; Parrett, D.F. Comparison of three weaning ages on cow-calf performance and steer carcass traits. *J. Anim. Sci.* **1999**, *77*, 323–329. [CrossRef]
21. Reddy, K.E.; Jeong, J.; Lee, S.D.; Baek, Y.C.; Oh, Y.; Kim, M.; So, K.M.; Kim, D.W.; Kim, J.H.; Park, S.; et al. Effect of different early weaning regimens for calves on adipogenic gene expression in Hanwoo loin at the fattening stage. *Livest. Sci.* **2017**, *195*, 87–98. [CrossRef]
22. Schoonmaker, J.P.; Cecava, M.J.; Faulkner, D.B.; Fluharty, F.L.; Zerby, H.N.; Loerch, S.C. Effect of source of energy and rate of growth on performance, carcass characteristics, ruminal fermentation, and serum glucose and insulin of early-weaned steers. *J. Anim. Sci.* **2003**, *81*, 843–855. [CrossRef] [PubMed]
23. Sithyphone, K.; Yabe, M.; Horita, H.; Hayashi, K.; Fumita, T.; Shiotsuka, Y.; Etoh, T.; Ebara, F.; Samadmanivong, O.; Wegner, J.; et al. Comparison of feeding systems: Feed cost, palatability and environmental impact among hay-fattened beef, consistent grass-only-fed beef and conventional marbled beef in Wagyu (Japanese Black cattle). *J. Anim. Sci.* **2011**, *82*, 352–359. [CrossRef]
24. Scheffler, J.M.; McCann, M.A.; Greiner, S.P.; Jiang, H.; Hanigan, M.D.; Bridges, G.A.; Gerrard, D.E. Early metabolic imprinting events increase marbling scores in fed cattle. *J. Anim. Sci.* **2014**, *92*, 320–324. [CrossRef] [PubMed]
25. Rasby, R. Early weaning beef calves. *Vet. Clin. N. Am. Food Anim. Pract.* **2007**, *23*, 29–40. [CrossRef]
26. Andrews, S. FastQC: A Quality Control Tool for High Throughput Sequence Data. Available online: http://www.bioinformatics.babraham.ac.uk/projects/fastqc/ (accessed on 2 June 2022).
27. Chen, S.; Zhou, Y.; Chen, Y.; Gu, J. Fastp: An ultra-fast all-in-one FASTQ preprocessor. *Bioinformatics* **2018**, *34*, i884–i890. [CrossRef] [PubMed]
28. Ewels, P.; Magnusson, M.; Lundin, S.; Käller, M. MultiQC: Summarize analysis results for multiple tools and samples in a single report. *Bioinformatics* **2016**, *32*, 3047–3048. [CrossRef] [PubMed]
29. Dobin, A.; Davis, C.A.; Schlesinger, F.; Drenkow, J.; Zaleski, C.; Jha, S.; Batut, P.; Chaisson, M.; Gingeras, T.R. STAR: Ultrafast universal RNA-seq aligner. *Bioinformatics* **2013**, *29*, 15–21. [CrossRef] [PubMed]
30. Liao, Y.; Smyth, G.K.; Shi, W. FeatureCounts: An efficient general purpose program for assigning sequence reads to genomic features. *Bioinformatics* **2014**, *30*, 923–930. [CrossRef]
31. Robinson, M.D.; McCarthy, D.J.; Smyth, G.K. EdgeR: A Bioconductor package for differential expression analysis of digital gene expression data. *Bioinformatics* **2010**, *26*, 139–140. [CrossRef]
32. R Core Team. R: A Language Land Environment for Statistical Computing. R Foundation for Statistical Computing. 2022. Vienna, Austria. Available online: https://www.R-project.org/ (accessed on 28 September 2022).
33. Kassambara, A.; Mundt, F. Factoextra: Extract and Visualize the Results of Multivariate Data Analyses. R Package Version 1.0. 7. 2020. Available online: https://CRAN.R-project.org/package=factoextra (accessed on 3 October 2022).
34. Wu, T.; Hu, E.; Xu, S.; Chen, M.; Guo, P.; Dai, Z.; Feng, T.; Zhou, L.; Tang, W.; Zhan, L.; et al. clusterProfiler 4.0: A universal enrichment tool for interpreting omics data. *Innovation* **2021**, *2*, 100141. [CrossRef] [PubMed]

35. Yu, G.; Wang, L.; Han, Y.; He, Q. clusterProfiler: An R package for comparing biological themes among gene clusters. *OMICS J. Int. Biol.* **2012**, *16*, 284–287. [CrossRef]
36. Kuleshov, M.V.; Jones, M.R.; Rouillard, A.D.; Fernandez, N.F.; Duan, Q.; Wang, Z.; Koplev, S.; Jenkins, S.L.; Jagodnik, K.M.; Lachmann, A.; et al. Enrichr: A comprehensive gene set enrichment analysis web server 2016 update. *Nucleic Acids Res.* **2016**, *44*, W90–W97. [CrossRef]
37. Myers, S.E.; Faulkner, D.B.; Nash, T.G.; Berger, L.L.; Parrett, D.F.; McKeith, F.K. Performance and carcass traits of early-weaned steers receiving either a pasture growing period or a finishing diet at weaning. *J. Anim. Sci.* **1999**, *77*, 311–322. [CrossRef] [PubMed]
38. Hulbert, L.E.; Moisá, S.J. Stress, immunity, and the management of calves. *J. Dairy Sci.* **2016**, *99*, 3199–3216. [CrossRef] [PubMed]
39. Meyer, D.L.; Kerley, M.S.; Walker, E.L.; Keisler, D.H.; Pierce, V.L.; Schmidt, T.B.; Stahl, C.A.; Linville, M.L.; Berg, E.P. Growth rate, body composition, and meat tenderness in early vs. traditionally weaned beef calves. *J. Anim. Sci.* **2005**, *83*, 2752–2761. [CrossRef] [PubMed]
40. Bonny, S.; Gardner, G.; Pethick, D.; Legrand, I.; Polkinghorne, R.; Hocquette, J. Biochemical measurements of beef are a good predictor of untrained consumer sensory scores across muscles. *Animal* **2015**, *9*, 179–190. [CrossRef]
41. Fonseca, L.F.S.; dos Santos Silva, D.B.; Gimenez, D.F.J.; Baldi, F.; Ferro, J.A.; Chardulo, L.A.L.; de Albuquerque, L.G. Gene expression profiling and identification of hub genes in Nellore cattle with different marbling score levels. *Genomics* **2020**, *112*, 873–879. [CrossRef]
42. Ladeira, M.M.; Schoonmaker, J.P.; Swanson, K.C.; Duckett, S.K.; Gionbelli, M.P.; Rodrigues, L.M.; Teixeira, P.D. Nutrigenomics of marbling and fatty acid profile in ruminant meat. *Animal* **2018**, *12*, s282–s294. [CrossRef]
43. Reggio, A.; Rosina, M.; Palma, A.; Cerquone Perpetuini, A.; Petrilli, L.L.; Gargioli, C.; Fuoco, C.; Micarelli, E.; Giuliani, G.; Cerretani, M.; et al. Adipogenesis of skeletal muscle fibro/adipogenic progenitors is affected by the WNT5a/GSK3/β-catenin axis. *Cell Death Differ.* **2020**, *10*, 2921–2941. [CrossRef]
44. Cianzo, D.S.; Topel, D.G.; Whitehurst, G.B.; Beitz, D.C.; Self, H.L. Adipose tissue growth and cellularity: Changes in bovine adipocyte and number. *J. Anim. Sci.* **1985**, *60*, 970–976. [CrossRef] [PubMed]
45. Wang, B.; Yang, Q.; Harris, C.L.; Nelson, M.L.; Busboom, J.R.; Zhu, M.J.; Du, M. Nutrigenomic regulation of adipose tissue development—Role of retinoic acid: A review. *Meat Sci.* **2016**, *120*, 100–106. [CrossRef] [PubMed]
46. Moisá, S.J.; Shike, D.W.; Faulkner, D.B.; Meteer, W.T.; Keisler, D.; Loor, J.J. Central role of the PPARγ gene network in coordinating beef cattle intramuscular adipogenesis in response to weaning age and nutrition. *Gene Regul. Syst. Biol.* **2014**, *8*, GRSB-S11782. [CrossRef] [PubMed]
47. Fu, Y.; Luo, N.; Klein, R.L.; Garvey, W.T. Adiponectin promotes adipocyte differentiation, insulin sensitivity, and lipid accumulation. *J. Lipid Res.* **2005**, *46*, 1369–1379. [CrossRef] [PubMed]
48. Wang, Y.H.; Bower, N.I.; Reverter, A.; Tan, S.H.; De Jager, N.; Wang, R.; McWilliam, S.M.; Cafe, L.M.; Greenwood, P.L.; Lehnert, S.A.; et al. Gene expression patterns during intramuscular fat development in cattle. *J. Anim. Sci.* **2009**, *87*, 119–130. [CrossRef] [PubMed]
49. Liu, S.; Huang, J.; Wang, X.; Ma, Y. Transcription factors regulate adipocyte differentiation in beef cattle. *Anim. Genet.* **2020**, *51*, 351–357. [CrossRef]
50. Graugnard, D.E.; Piantoni, P.; Bionaz, M.; Berger, L.L.; Faulkner, D.B.; Loor, J.J. Adipogenic and energy metabolism gene networks in *longissimus lumborum* during rapid post-weaning growth in Angus and Angus × Simmental cattle fed high-starch or low-starch diets. *BMC Genom.* **2009**, *10*, 142. [CrossRef]
51. Du, M.; Wang, B.; Fu, X.; Yang, Q.; Zhu, M.J. Fetal programming in meat production. *Meat Sci.* **2015**, *109*, 40–47. [CrossRef]
52. Smith, S.B.; Gill, C.A.; Lunt, D.K.; Brooks, M.A. Regulation of fat and fatty acid composition in beef cattle. *Asian-Australas. J. Anim. Sci.* **2009**, *22*, 1225–1233. [CrossRef]
53. Smith, S.B.; Blackmon, T.L.; Sawyer, J.E.; Miller, R.K.; Baber, J.R.; Morrill, J.C.; Wickersham, T.A. Glucose and acetate metabolism in bovine intramuscular and subcutaneous adipose tissues from steers infused with glucose, propionate, or acetate. *J. Anim. Sci.* **2018**, *96*, 921–929. [CrossRef]
54. Shijun, L.; Khan, R.; Raza SH, A.; Jieyun, H.; Chugang, M.; Kaster, N.; Linsen, Z. Function and characterization of the promoter region of perilipin 1 (PLIN1): Roles of E2F1, PLAG1, C/EBPβ, and SMAD3 in bovine adipocytes. *Genomics* **2020**, *112*, 2400–2409. [CrossRef]
55. Hudson, N.J.; Reverter, A.; Greenwood, P.L.; Guo, B.; Cafe, L.M.; Dalrymple, B.P. Longitudinal muscle gene expression patterns associated with differential intramuscular fat in cattle. *Animal* **2015**, *9*, 650–659. [CrossRef] [PubMed]
56. Du, M.; Yin, J.; Zhu, M.J. Cellular signaling pathways regulating the initial stage of adipogenesis and marbling of skeletal muscle. *Meat Sci.* **2010**, *86*, 103–109. [CrossRef] [PubMed]
57. Hossner, K.L. *Hormonal Regulation of Farm Animal Growth*; CABI: Wallingford, UK, 2005.
58. Zammit, P.S.; Relaix, F.; Nagata, Y.; Ruiz, A.P.; Collins, C.A.; Partridge, T.A.; Beauchamp, J.R. Pax7 and myogenic progression in skeletal muscle satellite cells. *J. Cell Sci.* **2006**, *119*, 1824–1832. [CrossRef] [PubMed]
59. Ustanina, S.; Carvajal, J.; Rigby, P.; Braun, T. The myogenic factor Myf5 supports efficient skeletal muscle regeneration by enabling transient myoblast amplification. *Stem Cells* **2007**, *25*, 2006–2016. [CrossRef]
60. Tapscott, S.J. The circuitry of a master switch: Myod and the regulation of skeletal muscle gene transcription. *J. Biol.* **2005**, *132*, 2685–2695. [CrossRef]

61. McPherron, A.C.; Lawler, A.M.; Lee, S.J. Regulation of skeletal muscle mass in mice by a new TGF-p superfamily member. *Nature* **1997**, *387*, 83–90. [CrossRef]
62. Guo, T.; Jou, W.; Chanturiya, T.; Portas, J.; Gavrilova, O.; McPherron, A.C. Myostatin inhibition in muscle, but not adipose tissue, decreases fat mass and improves insulin sensitivity. *PLoS ONE* **2009**, *4*, e4937. [CrossRef]
63. Kim, H.S.; Liang, L.; Dean, R.G.; Hausman, D.B.; Hartzell, D.L.; Baile, C. Inhibition of preadipocyte differentiation by myostatin treatment in 3T3-L1 cultures. *Biochem. Biophys. Res. Commun.* **2001**, *281*, 902–906. [CrossRef]
64. Obregon, M.J. Thyroid hormone and adipocyte differentiation. *Thyroid* **2008**, *18*, 185–195. [CrossRef]
65. Hausman, G.J.; Dodson, M.V.; Ajuwon, K.; Azain, M.; Barnes, K.M.; Guan, L.L.; Jiang, Z.; Poulos, S.P.; Sainz, R.D.; Smith, S.; et al. Board-invited review: The biology and regulation of preadipocytes and adipocytes in meat animals. *J. Anim. Sci.* **2009**, *87*, 1218–1246. [CrossRef]
66. Smith, S. The animal fatty acid synthase: One gene, one polypeptide, seven enzymes. *FASEB J.* **1994**, *8*, 1248–1259. [CrossRef] [PubMed]
67. Teixeira, P.D.; Oliveira, D.M.; Chizzotti, M.L.; Chalfun-Junior, A.; Coelho, T.C.; Gionbelli, M.; Paiva, L.V.; Carvalho, J.R.R.; Ladeira, M.M. Subspecies and diet affect the expression of genes involved in lipid metabolism and chemical composition of muscle in beef cattle. *Meat Sci.* **2017**, *133*, 110–118. [CrossRef] [PubMed]
68. Jiang, Z.; Michal, J.J.; Tobey, D.J.; Daniels, T.F.; Rule, D.C.; Macneil, M.D. Significant associations of stearoyl-CoA desaturase (SCD1) gene with fat deposition and composition in skeletal muscle. *Int. J. Biol. Sci.* **2008**, *4*, 345–351. [CrossRef] [PubMed]
69. Zhang, Y.; Zheng, Y.; Wang, X.; Qiu, J.; Liang, C.; Cheng, G.; Wang, H.; Zhao, C.; Yang, W.; Zan, L.; et al. Bovine Stearoyl-CoA desaturase 1 promotes adipogenesis by activating the PPARγ receptor. *J. Agric. Food Chem.* **2020**, *68*, 12058–12066. [CrossRef] [PubMed]
70. Jurie, C.; Cassar-Malek, I.; Bonnet, M.; Leroux, C.; Bauchart, D.; Boulesteix, P.; Pethick, D.W.; Hocquette, J.F. Adipocyte fatty acid-binding protein and mitochondrial enzyme activities in muscles as relevant indicators of marbling in cattle. *J. Anim. Sci.* **2007**, *85*, 2660–2669. [CrossRef]
71. Vasconcelos, J.T.; Sawyer, J.E.; Tedeschi, L.O.; McCollum, F.T.; Greene, L.W. Effects of different growing diets on performance, carcass characteristics, insulin sensitivity, and accretion of intramuscular and subcutaneous adipose tissue of feedlot cattle. *J. Anim. Sci.* **2009**, *87*, 1540–1547. [CrossRef]

metabolites

MDPI

Article

Chemical Characterization and In Vitro Gas Production Kinetics of Alternative Feed Resources for Small Ruminants in the Maltese Islands

Grazia Pastorelli [1,*], Kalliroi Simeonidis [1], Massimo Faustini [1,*], Angelo Le Mura [2], Mariagrazia Cavalleri [1,3], Valentina Serra [1] and Everaldo Attard [4]

1 Department of Veterinary Medicine and Animal Sciences, University of Milano, Via dell'Università 6, 26900 Lodi, Italy; kalliroi.simeonidis@studenti.unimi.it (K.S.); cavalleri@fbn-dummerstorf.de (M.C.); valentina.serra@unimi.it (V.S.)
2 Indipendent Researcher, 28053 Castelletto sopra Ticino, Italy; angelolemura@libero.it
3 Research-Institute for Farm Animal Biology (FBN), Institute of Nutritional Physiology, Wilhelm-Stahl-Allee 2, 18196 Dummerstorf, Germany
4 Division of Rural Sciences and Food Systems, Institute of Earth Systems, University of Malta, MSD 2080 Msida, Malta; everaldo.attard@um.edu.mt
* Correspondence: grazia.pastorelli@unimi.it (G.P.); massimo.faustini@unimi.it (M.F.); Tel.: +39-02-503-34576 (G.P.); +39-02-503-34569 (M.F.)

Abstract: The ever-increasing human population, the problem associated with climate change and recent crises—COVID-19 disease and trade conflicts—all impacted on the availability and cost of animal feed raw materials. This is clearly visible in realities which heavily rely on importation such as islands and small states, where producers involved in the agricultural sector were strongly affected by the sharp increase in prices. To deal with these global issues, alternative resources are perceived to replace conventional ingredients. This work aimed at assessing the nutritive value of different resources (sheep feed, mature carob, Maltese bread, wild asparagus, prickly lettuce, and loquat) for small ruminants present in the Maltese Islands, analyzing their chemical composition, gas production kinetics and antioxidant properties. In general, the variation in chemical composition resulted in different rumen fermentation kinetics ($p < 0.007$). The ratio between GP-24 h and GP-48 h was higher in Maltese bread than other substrates; loquat, prickly lettuce and wild asparagus showed lower fermentation kinetics in accordance with their high NDF and ADF contents. The antioxidant activity may be partially related to the polyphenolic content that was higher in wild asparagus, prickly lettuce and loquat. All feed characteristic confirmed their potential to be included as ingredients in ruminant diets and as a source of fiber.

Keywords: gas production technique; alternative feed; small ruminants; chemical analyses; antioxidant activity; polyphenols

1. Introduction

As a result of rising consumption, the world's natural resources are coming under increasing pressure from the ever-growing human population, which is projected to reach 10 billion by 2050 [1]. Additional factors to this scenario include the problems associated with climate change and recent crises, COVID-19 disease and trade conflicts, exacerbated existing food systems' vulnerabilities. Particularly, the global feed industry was faced with a situation that was never faced before: several external factors such as dangerous weather conditions in the growing season, as well as geopolitical issues, the impact of reliance on imports for staple foods, fodder for livestock, fertilizer and fuel impacted the availability and cost of feed raw material, which directly affected animal production. The impact is clearly visible in realities that heavily rely on importation, such as in the case of the Maltese Islands and other islands and small states which remain highly vulnerable to food price and foreign exchange volatility. To overcome the problems associated with the increase in the raw material prices, many farmers started to resort to alternative ingredients and new, potentially local sources to help meet low-cost formulation objectives.

Despite the small size of the Maltese Islands, they host a wide number of plants. Plant biodiversity, with its 1264 vascular species, is mainly attributed to the strategic position of Malta within the Mediterranean [2]. Some of these plants, present locally and in abundance, were considered as a potential alternative source for animal nutrition in order to mitigate the production costs of products of animal origin. Using natural vegetable resources may be an inexpensive, feasible and effective means to reduce production costs and will likely improve the quality of the products [3].

Although, in recent years, the interest in the use of alternative feed sources grew considerably, many of these were noy yet sufficiently investigated for their characteristics and properties. Therefore, there is a need for more information on the chemical composition and nutritive value of these resources to expand knowledge on their potential use as feedstuffs before making recommendations about their inclusion in animal diets. Chemical composition, in vitro digestibility and in vitro rumen fermentation kinetics are useful tools for feed evaluation, especially for ruminants. The in vitro techniques represent biological models that simulate the in vivo digestion processes with different levels of complexity. Compared to in vivo experiments, in vitro methods have the advantage not only of being less expensive, less time-consuming and more humane but allow for a more precise control of experimental conditions and for screening of a large number of materials in a relatively short time. The gas production test is an in vitro valuable tool in the evaluation and selection of feeds for ruminants, since it provides information on the fermentability, digestibility and nutritional value of feeds. Efficient fermentation implies a greater use of feed nutrients by rumen microorganisms, which translates into a higher conversion efficiency of feed into final products, such as volatile fatty acids and microbial protein. Thus, this contributes to a better diet formulation and an improvement in the feeding efficiency of ruminants. The aim of this study was to assess the nutritive value of alternative feeds resources for small ruminants in the Maltese Islands, analyzing their chemical composition, gas production kinetics and antioxidant properties.

2. Materials and Methods

2.1. Feed Resources

Five raw materials, Maltese bread, wild asparagus, loquat, prickly lettuce, mature carob and a commercial sheep feed, were used as substrates. Representative samples of each substrate, except for the sheep feed and Maltese bread, were collected from the rural area of the Maltese Islands. Maltese bread is a crusty sourdough bread, typical of the Maltese Islands. The parts of the feed sources tested are illustrated in Table 1. The sheep feed was composed of semolina, bran, corn, soya, alfalfa, sugarcane pulp, calcium carbonate, whereas the Maltese bread was made up of wheat flour, yeast, salt and water.

Table 1. Part of the feed resources tested.

Feed Resources	Scientific Name	Part Tested
Wild asparagus	*Asparagus aphyllus*	Aerial parts
Prickly lettuce	*Lactuca serriola*	Leaves
Loquat	*Eriobotrya japonica*	Leaves
Mature carob	*Ceratonia siliqua*	Pods
Maltese bread	*Ħobża Maltija* *	Crust and crumb

* Common name.

Analysis

All the samples were oven dried at 40 °C for 48 h and were then ground to pass a 1 mm sieve and stored for subsequent analysis prior to NIR analysis.

2.2. Proximate Analysis

Near infrared spectra of feed samples were acquired using a SpectraStar NIR spectrophotometer (the Unity® Scientific SpectraStarTM XT NIR Analyzer Series). Around 30 g of a bulk of feed samples were transferred to the quartz sample holder, sealed with a gold reflector, and placed over the sample window (1 cm of diameter), to ensure direct contact and minimize noises due to light scattering. Spectral data were acquired and recorded as absorbance spectra in wavelength range from 1400 to 2500 nm. During spectral acquisition, the sample was set to rotate and read three times (with a total of ninety determinations per sample) to ensure uniformity. All proximate analyses were conducted in triplicate. Parameters included dry matter (DM), crude protein (CP), ether extract (EE), neutral detergent fiber (NDF), acid detergent fiber (ADF) and total ash.

2.3. Inoculum Sources

The rumen fluid was collected from slaughtered sheep. In Malta, sheep breeds include the Comisana, the East Friesian, the Maltese and the Crossbreed. Rumen fluid was collected at a slaughterhouse from culled sheep previously fed under controlled conditions (i.e., commercial sheep feed and, additionally, wheat, sulla and clover), slaughtered in good health and transported from farms located near the slaughterhouse. The samples were collected from three sheep, which were (1 L per sheep) mixed and then treated with DMSO (5%) as cryoprotectant and delivered to laboratory within 30 min from slaughter. For the frozen preservation conditions, the tubes with 4 mL inocula containing a dimethyl sulfoxide (DMSO)/saline mixture were placed in a freezer at −80 °C to maintain inoculum homogeneity throughout the experiment.

2.4. In Vitro Gas Production

In vitro fermentation of the samples was conducted in an automated ANKOM gas production system (ANKOM RF; Ankom Technology Corp., New York, NY, USA). Briefly, buffered solution [4] was prepared and placed in a water-bath at 39 °C under continuous flushing with CO_2. Then, five jar bottles (250 mL) were filled with 78 mL of buffer solution along with 0.5 g of randomly assigned feedstuff (two of each feedstuff and one controls) per run. Four milliliters of inoculum were transferred to each fermentation bottle that, in turn, were immediately attached to the Ankom system. Blanks contained only the medium, inoculum and sheep feed. Bottles were incubated at 39 °C for 72 h. The pH of the contents of these bottles was recorded (pH meter, Thermo Scientific Orion 4-Star) at time 0 and at the end of the experimental period, that is, 72 h. The pH of the medium and CO_2 saturation at time 0 was controlled by a color change for the resazurin indicator from purple to pink/colorless. Pressure readings inside the bottles were recorded remotely every 10 min, through a wireless data logger system and the Ankom software (Gas Pressure Monitor, Ankom Technology Corp., Macedon, NY, USA) via a computer.

2.5. Physicochemical and Antioxidant Activity of Substrates Extracts

Prior to the assay, approximately 1 g of each dried subsample was allowed to macerate with 10 mL methanol in 50 mL centrifuge tubes. The supernatant was recovered and stored in the dark at −20 °C until analysis.

2.6. UV Analysis and Absorbance

The extracts (100 μL) were mixed with 900 μL of acidified methanol (1 M HCl:MeOH, 5:95) and scanned over a wavelength range of 200−800 nm in a UV-VIS spectrophotometer (Lightwave II, WPA). The following parameters were calculated:

$$\text{Tint Ratio} = (A_{420}/A_{520})$$
$$\text{Color intensity} = (A_{420} \times \text{DF}) + (A_{520} \times \text{DF}) + (A_{620} \times \text{DF})$$
$$\text{Flavonoid Ratio} = A_{420}/A_{520}$$
$$\text{Anthocyanin content (mg/kg)} = 1000 \times V_s \times \text{DF} \times A_{520}/\varepsilon$$

where A_{280}, A_{420}, A_{520}, and A_{620} are the absorbance values at 280, 420, 520 and 620 nm, respectively; DF is the dilution factor (sample to diluent = 100 µL:900 µL = 10); V_s is the volume of extract per gram of plant material; and ε is extinction coefficient [58.3 mL/(mg·cm)]. Results are the mean values (LSmean ± SEM) of three replicates.

2.7. Folin–Ciocalteu Test for Polyphenols/Total Phenolic Content

The total phenolic content (TPC) in methanol extracts was estimated using Folin–Ciocalteu (FC) reagent according to [5]. Gallic acid standard solution (960 µg/mL) was serially diluted 1:2 up to 60 µg/mL to construct the calibration curve. The FC reagent was diluted 1:10 with deionized water, while sodium carbonate was prepared as a 1 M solution. Briefly, 10 µL of extracts were pipetted in triplicate in wells of a 96-well microtiter plate (Nunc™ MicroWell™, ThermoFisher Scientific™,Waltham, MA, USA'). After adding 100 µL of FC reagent and 80 µL of sodium carbonate to each well, the plate was allowed to incubate at room temperature in the dark for 20 min and read at 750 nm on a spectrophotometer (SpectraMAX 340PC, Molecular Devices Corporation, San Jose, CA, USA). TPC was expressed as mg Gallic Acid Equivalents (GAE)/g. Results are the mean values (LSmean ± SEM) of three replicates.

2.8. Antioxidant Scavenging Assay

The DPPH free radical scavenging method is an antioxidant assay based on electron transfer which allow to calculate in a quick and simple way antioxidants by means of spectrophotometry. This assay was used as complementary method to evaluate the potential antioxidant activity by screening the DPPH free radical scavenging. The antioxidant activity of the substrates was determined using the microplate method [6].

Briefly, 100 µL aliquots of samples were transferred to wells in duplicates and in two sets. Subsequent one in two dilutions were carried out for these aliquots down the microtiter plate. One set was then treated with 100 µL methanol per well while the other set was treated with 100 µL DPPH solution per well. The plate was allowed to incubate at room temperature in the dark for 30 min and read at 517 nm in a microplate reader. Results were expressed as IC_{50} values. The values of IC_{50} denote the concentration of the sample that is required to scavenge 50% of DPPH free radicals. Results are the mean values (LSmean ± SEM) of three replicates.

2.9. Statistics Analysis

One-way ANOVA was performed with SPSS software (version 26.0 for Windows, SPSS Inc., Chicago, IL, USA). Duncan's test was applied to assess significant differences among the variables ($p < 0.05$) in terms of FC, DPPH and anthocyanins parameters. Principal component analysis was conducted on all feed for all physicochemical parameters studied, using XLSTAT (Microsoft, version 19.4.46756, SAS Institute Inc., Marlow, Buckinghamshire, UK) software. The gas produced during the fermentation process results in an increase in pressure measured over the chosen interval and was converted into gas volume using the ideal gas law. The data for cumulative GP (as mL of gas produced per g dry matter (DM) with time) for each profile were processed by the three-parameters Gompertz sigmoidal function. The choice of such function was decided by using the Akaike Information Criterion (AIC), an estimator of prediction error. The Gompertz function (three parameters) is defined as:

$$y(t) = a * e^{-b * e^{-c * t}}$$

where:

$y(t)$ is for the quantity of the gas produced at time t;

a = asymptotic value: estimated cumulative gas production;

b = latency time period;

c = growth rate;

e = Eulerian constant (e = 2.71828...). For each curve, the determination coefficient (R^2) was calculated.

For cumulative, latency and rate analysis of variance (ANOVA) model, considering substrate as fixed factor was applied. Multivariate hierarchical cluster analysis was performed using the chemical composition, and gas production kinetics data was analyzed to study emerging groupings of the feed. The method used for hierarchical agglomerative classification was complete linkage clustering based on a furthest neighbor criterion, with the furthest pair of observations between groups used to determine (dis)similarity of the groups. The similarity and dissimilarity measures were calculated as squared Euclidean distances.

3. Results

3.1. Proximate Analysis

The proximate analysis of feedstuffs is shown in Table 2. The chemical composition was quite variable between sources. DM varied in a ranging from 93.47 (wild asparagus) to 87.75 g/kg DM (mature carob). CP varied in a ranging from 14.20 (sheep feed) to 5.70 g/kg DM (loquat). EE content was also quite variable, with values from 6.43 (loquat) to 0.01 g/kg DM (mature carob). In the same way NDF content varied, from 56.22 (mature carob) to 36.80 g/kg DM (loquat). A large variation among sources was observed in the ADF content, ranging from 46.38 (wild asparagus) to 4.36 g/kg DM (Maltese bread). Additionally, ash widely varied with valued from 17.08 (loquat) to 5.27 g/kg DM (mature carob).

Table 2. Chemical composition (%) of individual feedstuffs. DM, dry matter; CP, crude protein; EE, ether extract; NDF, neutral detergent fiber; ADF, acid detergent fiber; ash content.

Feed Sources	DM	CP	EE	NDF	ADF	Ash
Sheep Feed	89.50 ± 0.04	14.2 ± 0.40	3.57 ± 0.03	46.64 ± 0.30	19.70 ± 0.40	9.70 ± 0.20
M_bread	90.4 ± 0.11	6.60 ± 0.15	2.90 ± 0.01	39.92 ± 0.42	4.36 ± 0.25	7.31 ± 0.17
W_asparagus	93.47 ± 0.14	8.90 ± 0.16	5.24 ± 0.18	50.45 ± 1.56	46.38 ± 0.53	12 ± 0.30
Pr_lettuce	90.05 ± 0.08	12.31 ± 0.18	2.31 ± 0.05	52.92 ± 0.38	38.64 ± 0.34	12.24 ± 0.27
Loquat	91.62 ± 0.09	5.70 ± 0.30	6.43 ± 0.05	36.80 ± 0.91	26.77 ± 0.58	17.08 ± 0.25
M_carob	87.75 ± 0	6.33 ± 0.07	0.01 ± 0	56.22 ± 0.32	25.80 ± 0.45	5.27 ± 0.03

3.2. In Vitro Gas Production and Fermentation Kinetics

Results from in vitro gas production kinetics parameters are shown in Table 3 and fitted total gas production is presented in Figure 1. The ratio of cumulative gas production at 24 h and 48 h (GP24/GP48) were compared in an attempt to ascertain how much of the fermentation was completed in the first 24 h [7]. Similar to the ratio of 48 h cumulative gas production and asymptotic gas production, a (GP48/a) were compared in order to determine how close 48 h gas production was from a. As with chemical composition, values from the evaluated parameters varied between substrates. The asymptotic gas production values (a) ranged from 74.62 (wild asparagus) to 183.43 mL/g DM (Maltese bread), which was significantly higher than all other substrates. Loquat had the numerical highest fractional rate of fermentation (c) value, while prickly lettuce had the lowest (0.066 h^{-1}). Regarding lag time (b), values varied from 2.32 h (loquat) to 4.17 h (wild asparagus). The final pH of substrates ranged between 6.8 and 6.9.

Table 3. a = asymptotic gas production (mL/g DM incubated); b = lag time (h); c = fractional rate of fermentation (h^{-1}); SEM = standard error of the mean; ns = not significant. ^{a,b,c} within column, means with different superscripts indicate statistical significance.

Substrate (S)	a	b	c	a24	a 48	24/48	48/a
Sheep feed	134.20 ^{bc}	2.87	0.080	85.41 ^{b}	121.25 ^{ab}	0.663 ^{b}	0.88 ^{ab}
Loquat	113.96 ^{ab}	2.32	0.210	71.098 ^{b}	99.75 ^{ab}	0.675 ^{b}	0.85 ^{ab}
M_carob	152.00 ^{bc}	2.48	0.078	100.47 ^{ab}	142.54 ^{a}	0.706 ^{ab}	1.00 ^{ab}
W_asparagus	74.62 ^{a}	4.17	0.080	45.86 ^{c}	66.32 ^{b}	0.691 ^{b}	0.88 ^{ab}
Pr_lettuce	111.29 ^{ab}	2.64	0.066	67.31 ^{b}	96.02 ^{ab}	0.700 ^{ab}	0.86 ^{ab}
M_bread	183.43 ^{c}	3.64	0.159	170.26 ^{a}	188.08 ^{ab}	0.960 ^{a}	1.01 ^{a}
SEM	8.55	0.26	0.03	9.87	10.5	0.002	0.03
P	0.007	ns	ns	<0.001	<0.001	0.005	0.016

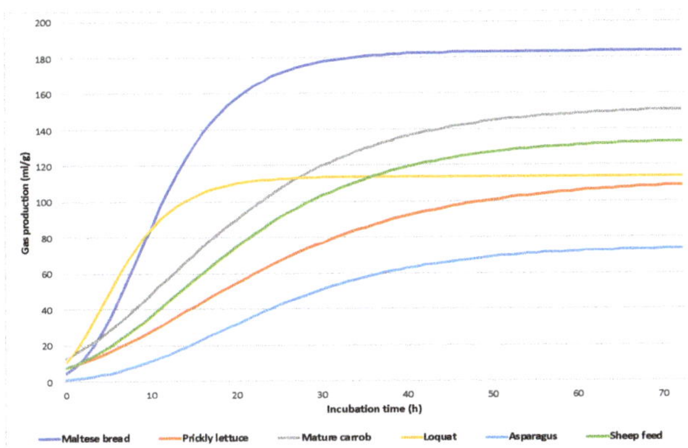

Figure 1. In vitro gas production (gas mL/g) of the tested substrates over time (h).

The cluster analysis using chemical composition data (Figure 2) showed a discrimination among substrates when data were used in a multivariate analysis. Cluster analysis detected the formation of two groups, the first being formed by wild asparagus and loquat (ashes and DM and EE) and the second being the rest of substrates, namely feed sheep, mature carob, bread, and prickly lettuce. A second analysis was performed on the fermentation kinetic data (Figure 3). Additionally, in this case, two clusters were identified; cluster A was singleton (wild asparagus) and B (others).

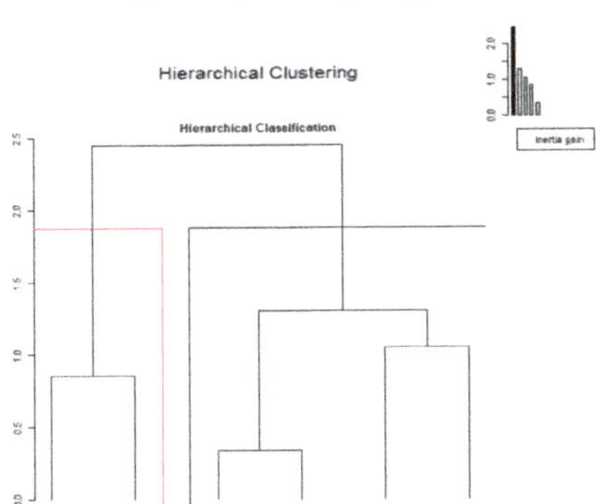

Figure 2. Hierarchical clustering of substrates using chemical composition.

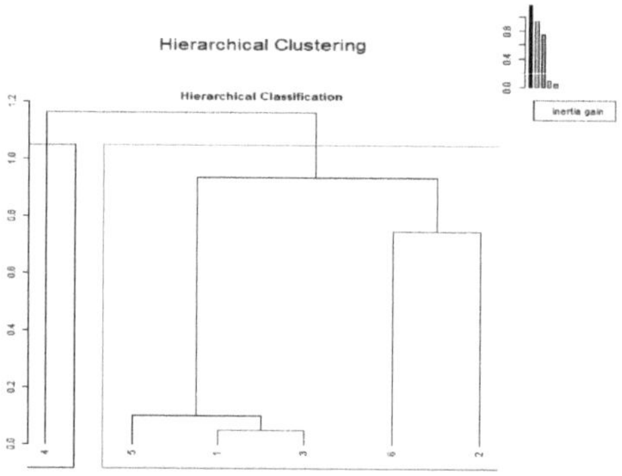

Figure 3. Hierarchical clustering of substrates using in vitro fermentation.

3.3. *UV-Vis Analysis*

Antioxidant Activity Assessment

The results for the physicochemical parameters (color index, tint ratio, and flavonoid ratio) and anthocyanin content are illustrated in Table 4 and Figure 4, respectively. The color index ranged between 0.3 A and 26.5 A and was exhibited most strongly by wild asparagus and prickly lettuce, whereas Maltese bread had the lowest value (0.301 A). The tint ratio ranged between 6.52 and 0.99, respectively for prickly lettuce and Maltese bread. The flavonoid ratio ranged between 0.02 and 0.23, respectively, for mature carob and Maltese bread. The anthocyanin measured by the spectrophotometric method showed values that differed significantly among feed sources as shown in Figure 5. Maltese bread was followed by mature carob and sheep feed, respectively, 1.7 ± 0.83, 4.75 ± 0.28 and 6.63 ± 0.47 mg/kg. The highest values were detected in wild asparagus and prickly lettuce, followed by loquat, respectively, 69.8 ± 5.7, 41.2 ± 2.4, and 16.6 ± 1.28 mg/kg. In detail, Maltese bread, mature carob, and sheep feed showed similar anthocyanin content (average value of 4.36 mg/kg), resulting significantly lower than loquat, prickly lettuce, and wild asparagus that showed the highest values settled around an average of 42.53 mg/kg.

Table 4. The color index (Abs), Tint ratio (A420/A520), and Flavonoid ratio (A520/A280) for each feed sources.

Feed Sources	Color Index (Abs)	Tint Ratio	Flavonoid Ratio
Sheep feed	3.02 ± 0.15	5.91 ± 0.23	0.05 ± 0.002
M_bread	0.3 ± 0.15	0.99 ± 0.013	0.23 ± 0.08
W_asparagus	26.5 ± 1.06	4.67 ± 0.39	0.09 ± 0.009
Pr_lettuce	20.2 ± 0.49	6.52 ± 0.44	0.07 ± 0.004
Loquat	3.55 ± 0.24	1.88 ± 0.041	0.03 ± 0.001
M_carob	1.04 ± 0.05	1.85 ± 0.045	0.02 ± 0.008

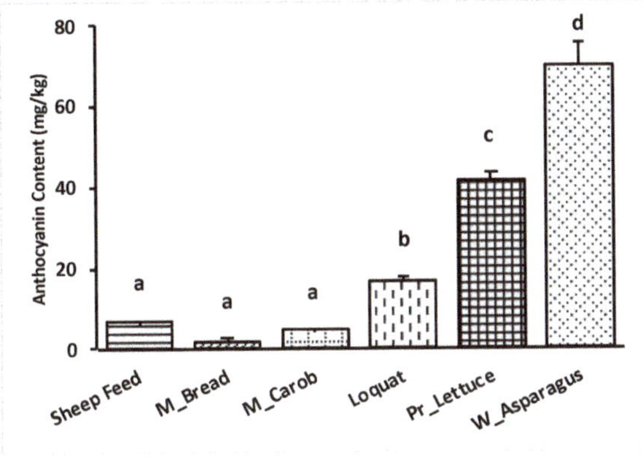

Figure 4. Anthocyanin content (mg/kg) of each feed sources. [abcd] Bars with different letters differ significantly at $p < 0.05$.

Figure 5 shows the UV-Vis profiles expressed as different wavelengths (nm) and relative absorbance (A) of each feed source. As shown in Table 5, the Maltese bread reported the highest values of red and blue flavonoids, respectively, 33 and 34.2%, while the highest values of yellow flavonoids were detected in prickly lettuce, sheep feed, and wild asparagus, respectively, 77.1, 75.4, and 71%. On the contrary, Maltese bread reported the lowest values of yellow flavonoids (32.8%), while the lowest values of red and blue flavonoids were detected in prickly lettuce (12 and 11%), sheep feed (13 and 12%), and wild asparagus (15.3 and 14%).

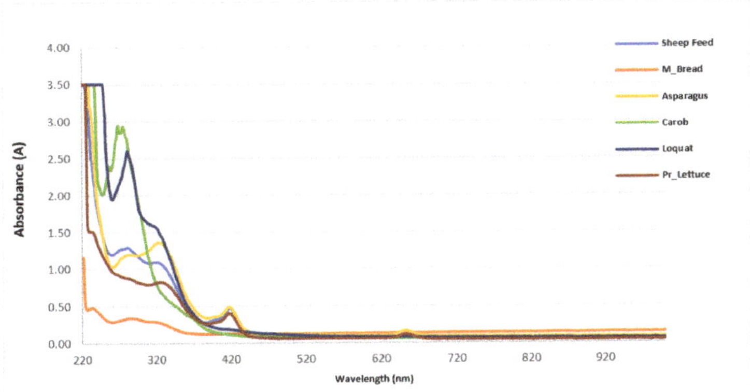

Figure 5. The UV-Vis profiles for each feed sources.

Table 5. Spearman correlation matrix for the physicochemical parameters of the feed sources.

Variables	Tint	%Red	%Yellow	%Blue	Antho	Flav Ratio	DM	Ash	CP	Fat	NDF	ADF	PolyP	DPPH
CI	0.714	−0.600	0.714	−0.714	1.000	0.086	0.543	0.714	0.257	0.429	0.143	0.943	0.771	−0.829
Tint		−0.943	1.000	−1.000	0.714	−0.029	−0.029	0.543	0.714	0.086	0.257	0.543	0.314	−0.486
%Red			−0.943	0.943	−0.600	0.086	0.257	−0.257	−0.771	0.200	−0.543	−0.486	−0.143	0.257
%Yellow				−1.000	0.714	−0.029	−0.029	0.543	0.714	0.086	0.257	0.543	0.314	−0.486
%Blue					−0.714	0.029	0.029	−0.543	−0.714	−0.086	−0.257	−0.543	−0.314	0.486
Antho						0.086	0.543	0.714	0.257	0.429	0.143	0.943	0.771	−0.829
Flav Ratio							0.543	0.029	0.371	0.143	−0.257	−0.029	−0.314	0.200
DM								0.600	−0.200	0.771	−0.543	0.486	0.543	−0.600
Ash									−0.029	0.657	−0.429	0.600	0.771	−0.886
CP										−0.143	0.257	0.029	−0.371	0.143
Fat											−0.771	0.257	0.543	−0.714
NDF												0.314	−0.086	0.257
ADF													0.829	−0.771
PolyP														−0.943

3.4. FC Test Results

The results for Folin–Ciocalteu assay are illustrated in Figure 6. The results are expressed as mg of Gallic acid equivalents (GAE). As shown, feed resources with the highest polyphenols content were loquat (4.4 mg GAE/g), which significantly differed from wild asparagus and prickly lettuce (3.3 mg GAE/g average value) and from mature carob (2.6 mg GAE/g) ($p < 0.05$). Sheep feed (1.26 mg GAE/g) and Maltese bread (0.52 mg GAE/g) were around an average of 0.9 mg GAE/g and were, hence, the lowest in average value.

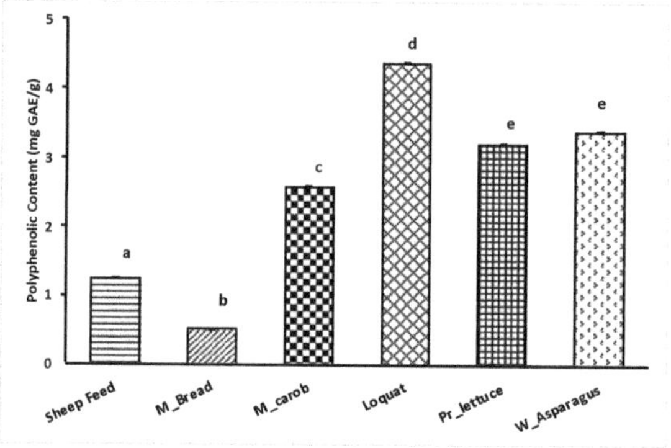

Figure 6. The polyphenolic content (mg GAE/g) of each feed source. [abcde] Bars with different letters differ significantly at $p < 0.05$.

3.5. DPPH Assay

The DPPH values are reported in Figure 7, showing the highest average values for Maltese bread (182 mg/mL) that differed significantly from mature carob (125.6 mg/mL) and the rest of the substrates. Loquat, wild asparagus, prickly lettuce, and sheep feed with the following content equal to 11.2, 11.4, 12.7, 19.2 mg/mL, respectively, indicated that they were more active than others ($p < 0.05$).

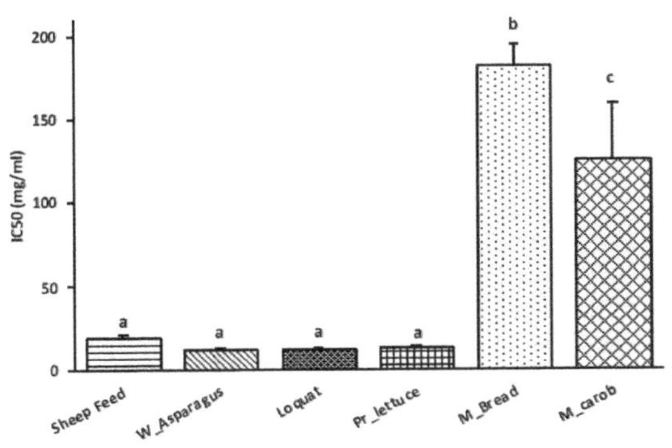

Figure 7. IC_{50} values (mg/mL) of each feed sources in relation to DPPH free radical scavenger. [abc] Bars with different letters differ significantly at $p < 0.05$.

4. Discussion

In the present study, we analyzed non-conventional feed resources (NCFR), except for sheep feed, which are widely distributed in the Maltese Islands and for this reason, their potential use in animal feeding raised interest mostly in local producers. NCFR generally refer to all those feeds that were not traditionally used for feeding livestock and are not commercially used in the production of livestock feed. Defined in this manner, NCFR can be looked at as covering a wide diversity of feeds and their nutrient contents. The feasibility of using NCFR depends on their feed value as well as their characteristics: the nutritive value is determined by the concentrations of its chemical components, as well as their rate and extent of digestibility. Thus, the efficient use of NCFR relies on their chemical and physical properties, which influence production system outputs. Since the scarcity of knowledge related to the chemical composition and nutritive value limits their use in feeding animals, it was important to analyze their chemical composition and fermentation kinetics.

As a general observation, chemical composition varied largely between the different sources analyzed. Variation in chemical composition could possibly be influenced by several factors, such as the variety, cultivation, harvesting, storage, and food processing techniques [8–11].

Wild asparagus reported higher values for both NDF and ADF related to the cultivated varieties [12]. Its greater fibrousness could be related by the harvesting time and determined by the micro-environment in which each spear grew and, in particular, by the prevalence of direct or diffused solar radiation [13]. High levels of NDF and ADF were found also in mature carob, loquat, and prickly lettuce. However, there are no existing data related to the chemical composition of loquat and prickly lettuce.

Concerning the prickly lettuce (*Lactuca serriola* L.), the main efforts were focused on morphological and, more recently, genetic characterization since the wild varieties represent potential genetic resources for improving nutritional traits in modern crop types [14]. For this reason, the metabolic aspects were barely taken into account, especially for the wild type. No data related to the content of NDF and ADF were found. However, crude protein, ether extract, ash content, and moisture content were in accordance with those reported by other authors [15].

Few studies observed differences in the aspects of chemical characterization of *Eriobotrya japonica* fruit and its parts (skin, pulp, seed, and starch) such as the vitamins, minerals, moisture, and proteins [16–18]. The chemical composition of the leaves present in our study is in accordance with the literature, except for ash content, which was higher than those reported by Hernández [19]. Ash is a reflection of the amount of mineral elements in the samples and, therefore, serves as the main source of mineral elements needed for human and animal health. Loquat leaves have a high amount of minerals, especially potassium, calcium, magnesium, iron, and sodium [20]. We speculated that the high content found in this study may depend on soil characteristics. The quantity and quality of ash depends on a large amount of factors including plant type, plant fraction, growing conditions, fertilization, choice of harvest date, harvest techniques, and conversion systems [21]. Carob (*Ceratonia siliqua* L.) is widely used as a source of fiber in animal feed in the Mediterranean region. The chemical composition of carob pods can vary with genetic, environmental, climatic factors, geographical origin, and harvesting season. The plant type and cultivar significantly influence chemical composition and biological activities of the carob pod [22]. Generally, it is characterized by high content of dietary fiber, hence classifying it as fibrous resource. Despite the NDF content being higher than those reported by Richane [23], NDF values of the studied carob samples were within the recommended range of 17–33 g/100 g DM for ruminants [23]. In general, carob is low in protein (3–4%) and lipid content (0.4–0.8%) [24]. The crude protein content in our study was above the minimum level required for maintenance of ruminants (6–8 g/100 g DM) [23] and was higher than those reported by Calislar and Kaplan [25] (4%) but similar to the one reported by Youssef [26] (6.3%). Nowadays, the main application of the pods is as animal feed. In fact, despite the limited information available concerning the chemical composition, nutritive values, and ruminal digestion of carob pods, it could represent a potential and economic alternative for grazing animals [27].

As expected, the wide range detected in chemical composition resulted in variability in rumen fermentation kinetics. In general, the fermentation parameters of feed varied according to the ingredients ($p = 0.007$). Additionally, there is little information on bread and no data related on the Maltese bread. Some evidence suggests that the chemical composition of the Maltese bread may be within the expected nutritional values [28]. An alternative feedstuff in feeding livestock is the bread by-product (BBP). Among the food industry by-products, bread waste from the bakery is an important source. The major restriction of feeding BBPs to livestock is the variation in the chemical composition [29]. However, some evidence suggested that diets including high levels of BBPs (30–45%) were shown to be better in vitro rumen degradation of starch, while showing a lower degradation of crude protein and fiber [30]. Maltese bread showed the highest volumes of gas production (a) and one of the highest rates of fermentation (c). These results indicated that Maltese bread fermented more extensively and at a faster rate than most of the other feed resources used in this study. Higher gas production indicates higher fermentative activity and higher nutrient degradation by rumen microorganisms. In fact, the carbohydrates' fractions differently affect the fermentation kinetics: starch promotes a more intense and rapid process, while conversely, structural carbohydrates cause a slower and less consistent fermentation, limiting the access to the cell content by micro-organisms, reducing nutrients' degradability, and slowing down the fermentation rate. The ratio between GP-24 h and GP-48 h was higher in Maltese bread than other substrates implying that a greater extent of fermentation had taken midway through the incubation period, showing a higher fermentation efficiency. This seemed to be confirmed by the fact that loquat, prickly lettuce, and especially wild asparagus showed the lowest volumes of gas production, indicating a lower fermentation kinetics. All these resources had a high content of NDF and ADF. The negative correlation between NDF level and gas production was consistent with other studies [31,32], in which a reduction in gas production with increasing of NDF content of the substrates was reported. In general, the present results were lower than those found in the literature [33,34]. Aderinboye et al. [33] found that cumulative gas production and the rate of fermentation were generally higher with cattle than sheep ruminal fluid. The same result was consistent with Cone et al. [34]; however, the activity of the rumen fluid is not only determined by the donor animal, but also by the ration of the animal.

As was highlighted by most authors and reviews [35], the composition and type of diet given to the donor animals determined the fermentative capacity and the degree of the adaptation of the analyzed substrates. Thus, the rate and extent of fermentation of a fibrous or a concentrate feed highly depended on its evaluation with inoculum from an animal feed on either a forage- or a concentrate-type diet [36]. Therefore, it is recommended to feed donor animals with a diet similar to the substrate to be incubated in vitro, or to the in vivo feeding conditions that are intended to be studied. In fact, in the present paper, these conditions were met. Rather, we hypothesized that such a result can also be due to the storage (frozen inocula) that could have affected the microbial activity declining over time and influencing the fermentation kinetics of the substrates [37,38]. Lag time is indicative of the time taken for microbes to adhere themselves to the substrates, and microbial attachment to insoluble substrate. In the present study, no significant difference was found in c parameter to underline a homogeneous pre-condition for digestion to proceed. However, the main advantage of using frozen inocula remains justified, since the use of fistulated animals is avoided, aliquots can be prepared and used for more than one fermentation and results are repeatable as highlighted in our study. Moreover, results of gas production obtained from different researchers in vitro are often, in any case, difficult to compare because of the influence of several confounding sources of variation: the procedures of rumen fluid collection and treatment [39], the type of buffer used [40], the ratios among feed sample size, fermentation fluid, headspace volume [41], and the type of GP equipment, such as syringes or bottles that can be closed or vented at fixed times or at fixed pressure [42–44]. Considering sheep feed as a control since it is a reference feed in sheep nutrition, with regard to the cumulative gas production, we observed that other feeds, except for wild asparagus, behave in a similar way, suggesting a substrate interchangeability.

In the present study, all the feed resources were evaluated for their content of bioactive compounds as phenolic contents and for their antioxidant activity. The exogenous antioxidants were mainly derived from food and medicinal plants, such as fruits, vegetables, cereals, mushrooms, beverages, flowers, spices, and traditional medicinal herbs [45,46]. It was found that phenolic compounds are amongst the most effective antioxidant constituents in plant foods, including fruits, vegetables, and grains [47]. They are a ubiquitous group of phytochemicals that possess different physiological activities associated with their chemical structures. Considering their important health effects, the efficient extraction methods of natural antioxidants, appropriate assessment of antioxidant activity as well as their main resources from food and medicinal plants drew great attention in food science and nutrition.

Although carob is widely used as a source of fiber in animal feed, it holds potentially significant importance due to its phytochemical constituents with functional properties, flavoring properties, and nutrition benefits [48]. In the present study, the polyphenol content of mature carob was in accordance with the literature, while the anthocyanin content showed lower values [23]. The anthocyanin content may be correlated to irradiance levels at which the plants were grown [49]. The antioxidant activity was lower than what was reported in the literature. In fact, it was demonstrated that the antioxidant activity decreased significantly throughout the ripening process [50]. For this reason, the maturity of the carob may have influenced the antioxidant power of this phytochemical group.

Several studies reported different bioactive compounds, associated with loquat health-relevant properties, making its use in food industry desirable. Bioactivity was observed in various parts of the plant and its fruit, including the leaves [51]. In leaves, the antioxidant capacity is linked to the phenolic compounds [52]. Some authors [46,53–55] demonstrated a linear correlation between the content of total phenolic compounds and their antioxidant capacity. Our results showed a similar polyphenolic content compared with that reported by other authors [56]. The low IC_{50} value found herein and flavonoid content indicated a valid antioxidant activity in loquat leaves.

Many studies investigated the diversity and abundance of polyphenols in different types and cultivars of asparagus under various growing conditions [57,58]. The content of polyphenols in our study was similar to that reported by di Maro [59], but lower than those reported by other authors [60,61]. The content of anthocyanins was lower than those reported by Dong et al. [62]. The DPPH scavenging activity was similar to that reported by Adouni et al. [61], but lower than that reported in fresh spears [59].

Although several plants of the *Lactuca* genus were examined for their chemical constituents, prickly lettuce was not investigated in detail for its phytochemical constitution. We reported a high content of anthocyanins (41.2 mg/mL), compared to the literature [63,64]. The DPPH scavenging activity was in accordance with the literature [65], indicating a valid antioxidant activity in prickly lettuce. In general, mature carob, wild asparagus, prickly lettuce, and loquat could be potentially used as a source of fiber and with reference to the last three feeds, our data suggested that the highest DPPH radical scavenging activity make them also as valid antioxidant feeds.

Additionally, there were no data related on the Maltese bread. The antioxidant activity of bread is usually attributed to the presence of Maillard reaction products that are known to possess free-radical scavenging activities [66]. In this case, lower antioxidant activity was noted for the Maltese bread in comparison to the other substrates. This could be correlated with the baking process. In fact, during bread baking, various modifications in the chemical composition and properties of the food matrix take place, leading to changes in the nutritional value of the final product [67]. This treatment can lead to the production of newly formed compounds responsible for different biological activities [68].

Finally, the antioxidant power of sheep feed was also evaluated. The lower IC_{50} value found herein indicated a valid antioxidant activity in sheep feed. This could be correlated with the presence of different type of wheat (semolina, bran, corn), and alfalfa which are rich sources of antioxidant flavonoids and phenolic compounds reported to have anti-inflammatory and antioxidant activity [69,70]. Additionally, it was previously reported that different varieties of sugarcane show good antioxidant properties [71]. In fact, the antioxidant activity is susceptible to variation among varieties, growing practices, processing and storage conditions on the biologically active compounds.

Principal component analysis was conducted on all six feeds to determine any latent differences between them, taking into account all the physicochemical characteristics under study. Spearman correlation (Table 5) revealed that there were strong positive correlations between color intensity, tint, % yellow, anthocyanin content, ash, ADF, and polyphenols (r_s = 0.714–1.000), indicating that the physical parameters correlate with the chemical constitution of these feeds. Additionally, anthocyanins and polyphenols exhibited a positive correlation (r_s = 0.771) but a negative correlation between anthocyanins and DPPH (r_s = −0.829), indicating that anthocyanins contributed to the total polyphenolic content whilst also contributing to the antioxidant activity of the feeds.

Additionally, DPPH correlated negatively with most parameters ($r_s < -0.486$), indicating that most feed components (fiber, fat, and polyphenols amongst others) contribute to the antioxidant activity of these natural sources. Two latent factors had an eigenvalue greater than 1, which together explained 77.82% of the total variance (Figure 8a). The factor loadings demonstrated the different groups of variables. The first factor was loaded mostly on physical parameters, polyphenolic content, and antioxidant activity, while the second factor was loaded on most proximate parameters. Figure 8b demonstrates the factor scores of the two latent factors. Factor 1, on the horizontal axis, demonstrates the clustering of leafy feed sources (loquat, wild asparagus, and prickly lettuce) on the right-hand side of the scatter plot, with the rest more shifted towards the right. Factor 2, on the vertical axis, demonstrates the clustering of feeds, at the bottom of the plot, with low content of dry matter and fat, whilst a high content of protein and fiber. These demonstrate that although the leafy feed sources are distinctive from the rest, they may share some common properties with other feeds, indicating that all feed sources may provide benefit to ruminants in one way or another.

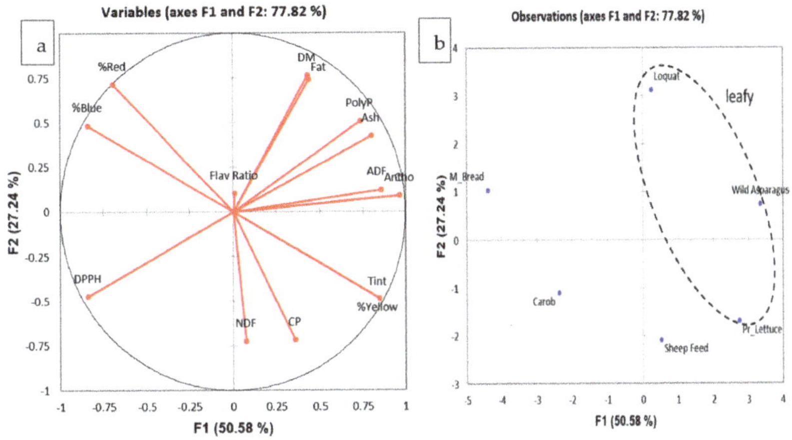

Figure 8. Principal component analysis (PCA) analysis of the six feed sources for the physicochemical, proximate and antioxidant parameters; (**a**) the factor loading plot demonstrating the different groups of variables; (**b**) the factor scores of the two latent factors.

5. Conclusions

This study provided a chemical, antioxidant, and kinetics evaluation of different feed resource, hence giving a first insight in the characterization of some local feeds such as wild asparagus, pricky lettuce, and loquat. The results of this first investigation suggest the possibility of using local feeds in small ruminant nutrition with the advantage that, being local natural resources, they are better adapted to the climate and agronomic conditions and limit the environmental impact. Moreover, further research will be needed in order to establish the right amounts of the feedstuffs to be added in the diets of livestock based on animal production responses and, hence, their economic value.

Author Contributions: Conceptualization, G.P. and E.A.; methodology, G.P., K.S., E.A., A.L.M. and M.C.; software, M.F., E.A. and G.P.; validation, G.P., E.A. and M.F.; formal analysis, E.A., K.S., A.L.M. and M.C.; investigation, E.A., G.P., K.S., E.A., M.C., A.L.M. and V.S.; data curation, G.P., M.F. and E.A.; writing—original draft preparation, G.P., K.S., E.A., M.F. and V.S.; writing—review and editing, G.P., K.S., V.S. and E.A.; visualization, G.P., K.S., M.F., A.L.M., M.C., V.S. and E.A.; supervision, G.P. and E.A.; funding acquisition, E.A. All authors have read and agreed to the published version of the manuscript.

Funding: This research received no external funding. APC was funded thanks to IOAP discount of University of Milano and Author Voucher discount.

Institutional Review Board Statement: Not applicable.

Informed Consent Statement: Not applicable.

Data Availability Statement: The data presented in the current study are available from the corresponding authors on reasonable request due to privacy or ethical restrictions.

Conflicts of Interest: The authors declare no conflict of interest.

References

1. United Nations (UN). *World Population Prospects: The 2017 Revision, Key Findings and Advance Tables*; United Nations: New York, NY, USA, 2019.
2. Attard, E.; Pacioni, P. The Phytochemical and In Vitro Pharmacological Testing of Maltese Medicinal Plants. In *Bioactive Compounds in Phytomedicine*; InTech: Rijeka, Croatia, 2012; pp. 93–112.
3. Estévez, M. Critical overview of the use of plant antioxidants in the meat industry: Opportunities, innovative applications and future perspectives. *Meat Sci.* **2021**, *181*, 108610. [CrossRef]

4. Menke, K.H. Estimation of the energetic feed value obtained from chemical analysis and in vitro gas production using rumen fluid. *Anim. Res. Dev.* **1988**, *28*, 7–55.
5. Attard, E. A rapid microtitre plate Folin-Ciocalteu method for the assessment of polyphenols. *Open Life Sci.* **2013**, *8*, 48–53. [CrossRef]
6. Prieto, J.M. Procedure: Preparation of DPPH Radical, and antioxidant scavenging assay. *DPPH Microp. Protoc.* **2012**, 7–9.
7. Bueno, I.C.S.; Cabral Filho, S.L.S.; Gobbo, S.P.; Louvandini, H.; Vitti, D.M.S.S.; Abdalla, A.L. Influence of inoculum source in a gas production method. *Anim. Feed Sci. Technol.* **2005**, *123*, 95–105. [CrossRef]
8. Chitrakar, B.; Zhang, M.; Adhikari, B. Asparagus (*Asparagus officinalis*): Processing effect on nutritional and phytochemical composition of spear and hard-stem byproducts. *Trends Food Sci. Technol.* **2019**, *93*, 1–11. [CrossRef]
9. Drinkwater, J.M.; Tsao, R.; Liu, R.; Defelice, C.; Wolyn, D.J. Effects of cooking on rutin and glutathione concentrations and antioxidant activity of green asparagus (*Asparagus officinalis*) spears. *J. Funct. Foods* **2015**, *12*, 342–353. [CrossRef]
10. Gonnella, M.; Durante, M.; Caretto, S.; D'Imperio, M.; Renna, M. Quality assessment of ready-to-eat asparagus spears as affected by conventional and sous-vide cooking methods. *LWT* **2018**, *92*, 161–168. [CrossRef]
11. Liu, Y.; Zhang, W.; Xu, C.; Li, X. Biological Activities of Extracts from Loquat (*Eriobotrya japonica* Lindl.): A Review. *Int. J. Mol. Sci.* **2016**, *17*, 1983. [CrossRef] [PubMed]
12. Sergio, L.; Boari, F.; Di Venere, D.; Gonnella, M.; Cantore, V.; Renna, M. Quality Evaluation of Wild and Cultivated Asparagus: A Comparison between Raw and Steamed Spears. *Agriculture* **2021**, *11*, 1213. [CrossRef]
13. Maeda, T.; Honda, K.; Sonoda, T.; Motoki, S.; Inoue, K.; Suzuki, T.; Oosawa, K.; Suzuki, M. Light Condition Influences Rutin and Polyphenol Contents in Asparagus Spears in the Mother-fern Culture System during the Summer–Autumn Harvest. *J. Jpn. Soc. Hortic. Sci.* **2010**, *79*, 161–167. [CrossRef]
14. Lebeda, A.; Křístková, E.; Kitner, M.; Mieslerová, B.; Jemelková, M.; Pink, D.A.C. Resistance of wild *Lactuca* genetic resources to diseases and pests, and their exploitation in lettuce breeding. *Acta Hortic.* **2015**, *1101*, 133–140. [CrossRef]
15. Kim, S.W.; Huh, W.; Kim, S.Y.; Kim, J.S.; Lee, K.J. Isolation and Characterization of Bio-active Materials from Prickly Lettuce (*Lactuca serriola*). Sung-Il Moon. *J. Life Sci.* **2009**, *19*, 206–212.
16. Costa, B.P.; Carpiné, D.; da Silva Bambirra Alves, F.E.; Barbi, R.C.T.; de Melo, A.M.; Ikeda, M.; Ribani, R.H. Thermal, structural, morphological and bioactive characterization of acid and neutral modified loquat (*Eriobotrya japonica* Lindl.) seed starch and its by-products. *J. Therm. Anal. Calorim.* **2022**, *147*, 6721–6737. [CrossRef]
17. Dhiman, A.; Suhag, R.; Thakur, D.; Gupta, V.; Prabhakar, P.K. Current Status of Loquat (Eriobotrya Japonica Lindl.): Bioactive Functions, Preservation Approaches, and Processed Products. *Food Rev. Int.* **2022**, *38*, 286–316. [CrossRef]
18. Turola Barbi, R.C.; Teixeira, G.L.; Hornung, P.S.; Ávila, S.; Hoffmann-Ribani, R. *Eriobotrya japonica* seed as a new source of starch: Assessment of phenolic compounds, antioxidant activity, thermal, rheological and morphological properties. *Food Hydrocoll.* **2018**, *77*, 646–658. [CrossRef]
19. Hernández, F.; Madrid, J.; Cerón, J.J.; Pulgar, M.A.; Cid, J.M. Utilisation of lemon (*Citrus limon*) and loquat (*Eribotrya japonica*) tree leaves alone or with NH3-treated straw for goats. *J. Sci. Food Agric.* **1998**, *77*, 133–139. [CrossRef]
20. Khouya, T.; Ramchoun, M.; Elbouny, H.; Hmidani, A.; Bouhlali, E.d.T.; Alem, C. Loquat (*Eriobotrya japonica* (Thunb) Lindl.): Evaluation of nutritional value, polyphenol composition, antidiabetic effect, and toxicity of leaf aqueous extract. *J. Ethnopharmacol.* **2022**, *296*, 115473. [CrossRef] [PubMed]
21. Bakker, R.R.; Elbersen, H.W. Managing ash content and quality in herbaceous biomass: An analysis from plant to product managing ash content and-quality in herbaceous biomass: An analysis from plant to product. In Proceedings of the 14th European Biomass Conference, Paris, France, 17–21 October 2005; p. 21.
22. Simsek, S.; Ozcan, M.M.; Al Juhaimi, F.; ElBabiker, E.; Ghafoor, K. Amino Acid and Sugar Contents of Wild and Cultivated Carob (*Ceratonia siliqua*) Pods Collected in Different Harvest Periods. *Chem. Nat. Compd.* **2017**, *53*, 1008–1009. [CrossRef]
23. Richane, A.; Ismail, H.B.; Darej, C.; Attia, K.; Moujahed, N. Potential of Tunisian carob pulp as feed for ruminants: Chemical composition and in vitro assessment. *Trop. Anim. Health Prod.* **2022**, *54*, 58. [CrossRef]
24. Kumazawa, S.; Taniguchi, M.; Suzuki, Y.; Shimura, M.; Kwon, M.S.; Nakayama, T. Antioxidant Activity of Polyphenols in Carob Pods. *J. Agric. Food Chem.* **2002**, *50*, 373–377. [CrossRef]
25. Calislar, S.; Kaplan, Y. Effects of carob (*Ceratonia siliqua*) pod byproduct on quail performance, egg characteristics, fatty acids, and cholesterol levels. *Rev. Bras. Zootec.* **2017**, *46*, 113–117. [CrossRef]
26. Youssef, M.K.E.; El-Manfaloty, M.M.; Ali, H.M. Assessment of proximate chemical composition, nutritional status, fatty acid composition and phenolic compounds of carob (*Ceratonia siliqua* L.). *Food Public Health* **2013**, *3*, 304–308.
27. Kaltenegger, A.; Humer, E.; Stauder, A.; Zebeli, Q. Feeding of bakery by-products in the replacement of grains enhanced milk performance, modulated blood metabolic profile, and lowered the risk of rumen acidosis in dairy cows. *J. Dairy Sci.* **2020**, *103*, 10122–10135. [CrossRef] [PubMed]
28. Pinotti, L.; Luciano, A.; Ottoboni, M.; Manoni, M.; Ferrari, L.; Marchis, D.; Tretola, M. Recycling food leftovers in feed as opportunity to increase the sustainability of livestock production. *J. Clean. Prod.* **2021**, *294*, 126290. [CrossRef]
29. Guiroy, P.J.; Fox, D.G.; Beermann, D.H.; Ketchen, D.J. Performance and meat quality of beef steers fed corn-based or bread by-product-based diets. *J. Anim. Sci.* **2000**, *78*, 784. [CrossRef] [PubMed]

30. Humer, E.; Aditya, S.; Kaltenegger, A.; Klevenhusen, F.; Petri, R.M.; Zebeli, Q. Graded substitution of grains with bakery by-products modulates ruminal fermentation, nutrient degradation, and microbial community composition in vitro. *J. Dairy Sci.* **2018**, *101*, 3085–3098. [CrossRef]
31. Chino Velasquez, L.B.; Molina-Botero, I.C.; Moscoso Muñoz, J.E.; Gómez Bravo, C. Relationship between Chemical Composition and In Vitro Methane Production of High Andean Grasses. *Animals* **2022**, *12*, 2348. [CrossRef] [PubMed]
32. Váradyová, Z.; Baran, M.; Zeleňák, I. Comparison of two in vitro fermentation gas production methods using both rumen fluid and faecal inoculum from sheep. *Anim. Feed. Sci. Technol.* **2005**, *123*, 81–94. [CrossRef]
33. Aderinboye, R.Y.; Akinlolu, A.O.; Adeleke, M.A.; Najeem, G.O.; Ojo, V.O.A.; Isah, O.A.; Babayemi, O.J. In vitro gas production and dry matter degradation of four browse leaves using cattle, sheep and goat inocula. *Slovak J. Anim. Sci.* **2016**, *49*, 32–43.
34. Cone, J.W.; van Gelder, A.H.; Bachmann, H. Influence of inoculum source on gas production profiles. *Anim. Feed Sci. Technol.* **2002**, *99*, 221–231. [CrossRef]
35. Mould, F.L.; Morgan, R.; Kliem, K.E.; Krystallidou, E. A review and simplification of the in vitro incubation medium. *Anim. Feed Sci. Technol.* **2005**, *123*, 155–172. [CrossRef]
36. Amanzougarene, Z.; Yuste, S.; Fondevila, M. Fermentation Pattern of Several Carbohydrate Sources Incubated in An in Vitro Semicontinuous System with Inocula from Ruminants Given Either Forage or Concentrate-Based Diets. *Animals* **2020**, *10*, 261. [CrossRef]
37. Borba, A.E.S.; Correia, P.J.A.; Fernandes, J.M.M.; Borba, A.F.R.S. Comparison of three sources of inocula for predicting apparent digestibility of ruminant feedstuffs. *Anim. Res.* **2001**, *50*, 265–273. [CrossRef]
38. Chaudhry, A.S.; Mohamed, R.A.I. Fresh or frozen rumen contents from slaughtered cattle to estimate in vitro degradation of two contrasting feeds. *Czech J. Anim. Sci.* **2012**, *57*, 265–273. [CrossRef]
39. Cornou, C.; Storm, I.M.L.D.; Hindrichsen, I.K.; Worgan, H.; Bakewell, E.; Ruiz, D.R.; Yáñez, A.L.; Tagliapietra, F.; Cattani, M.; Ritz, C.; et al. A ring test of a wireless in vitro gas production system. *Anim. Prod. Sci.* **2013**, *53*, 585. [CrossRef]
40. Patra, A.K.; Yu, Z. Effects of gas composition in headspace and bicarbonate concentrations in media on gas and methane production, degradability, and rumen fermentation using in vitro gas production techniques. *J. Dairy Sci.* **2013**, *96*, 4592–4600. [CrossRef]
41. Ramin, M.; Huhtanen, P. Development of an in vitro method for determination of methane production kinetics using a fully automated in vitro gas system—A modelling approach. *Anim. Feed. Sci. Technol.* **2012**, *174*, 190–200. [CrossRef]
42. Gierus, M.; Schiborra, A.; Südekum, K.H.; Rave, G.; Taube, F. Comparison of gas accumulation profiles of several feeds using manual or automated gas production methods. *Anim. Feed Sci. Technol.* **2008**, *147*, 310–325. [CrossRef]
43. Pirondini, M.; Malagutti, L.; Colombini, S.; Amodeo, P.; Crovetto, G.M. Methane yield from dry and lactating cows diets in the Po Plain (Italy) using an in vitro gas production technique. *Ital. J. Anim. Sci.* **2012**, *11*, e61. [CrossRef]
44. Tagliapietra, F.; Cattani, M.; Hansen, H.H.; Bittante, G.; Schiavon, S. High doses of vitamin E and vitamin C influence in vitro rumen microbial activity. *Anim. Feed. Sci. Technol.* **2013**, *183*, 210–214. [CrossRef]
45. Zhang, J.J.; Li, Y.; Zhou, T.; Xu, D.P.; Zhang, P.; Li, S.; Li, H.B. Bioactivities and Health Benefits of Mushrooms Mainly from China. *Molecules* **2016**, *21*, 938. [CrossRef] [PubMed]
46. Cai, Y.; Luo, Q.; Sun, M.; Corke, H. Antioxidant activity and phenolic compounds of 112 traditional Chinese medicinal plants associated with anticancer. *Life Sci.* **2004**, *74*, 2157–2184. [CrossRef]
47. Tsao, R. Chemistry and Biochemistry of Dietary Polyphenols. *Nutrients* **2010**, *2*, 1231–1246. [CrossRef]
48. Papaefstathiou, E.; Agapiou, A.; Giannopoulos, S.; Kokkinofta, R. Nutritional characterization of carobs and traditional carob products. *Food Sci. Nutr.* **2018**, *6*, 2151–2161. [CrossRef]
49. Vinterhalter, B.; Ninkovic, S.; Kozomara, B.; Vinterhalter, D. Carbohydrate nutrition and anthocyanin accumulation in light grown and etiolated shoot cultures of carob (*Ceratonia siliqua* L.). *Arch. Biol. Sci.* **2007**, *59*, 51–56. [CrossRef]
50. Benchikh, Y.; Louaileche, H.; George, B.; Merlin, A. Changes in bioactive phytochemical content and in vitro antioxidant activity of carob (*Ceratonia siliqua* L.) as influenced by fruit ripening. *Ind. Crops Prod.* **2014**, *60*, 298–303. [CrossRef]
51. Wu, Q.L.; Wang, M.; Simon, J.E.; Yu, S.C.; Xiao, P.G.; Ho, C.T. Studies on the Chemical Constituents of Loquat Leaves (*Eriobotrya japonica*). In *Oriental Foods and Herbs*; ACS Symposium Series; ACS Publications: Washington, DC, USA, 2003; pp. 292–306.
52. Ferreres, F.; Gomes, D.; Valentão, P.; Gonçalves, R.; Pio, R.; Chagas, E.A.; Seabra, R.M.; Andrade, P.B. Improved loquat (*Eriobotrya japonica* Lindl.) cultivars: Variation of phenolics and anti- oxidative potential. *Food Chem.* **2009**, *114*, 1019–1027. [CrossRef]
53. Djeridane, A.; Yousfi, M.; Nadjemi, B.; Boutassouna, D.; Stocker, P.; Vidal, N. Antioxidant activity of some algerian medicinal plants extracts containing phenolic compounds. *Food Chem.* **2006**, *97*, 654–660. [CrossRef]
54. Katalinic, V.; Milos, M.; Kulisic, T.; Jukic, M. Screening of 70 medicinal plant extracts for antioxidant capacity and total phenols. *Food Chem.* **2006**, *94*, 550–557. [CrossRef]
55. Katsube, T.; Tabata, H.; Ohta, Y.; Yamasaki, Y.; Anuurad, E.; Shiwaku, K.; Yamane, Y. Screening for Antioxidant Activity in Edible Plant Products: Comparison of Low-Density Lipoprotein Oxidation Assay, DPPH Radical Scavenging Assay, and Folin−Ciocalteu Assay. *J. Agric. Food Chem.* **2004**, *52*, 2391–2396. [CrossRef]
56. Gruz, J.; Ayaz, F.A.; Torun, H.; Strnad, M. Phenolic acid content and radical scavenging activity of extracts from medlar (*Mespilus germanica* L.) fruit at different stages of ripening. *Food Chem.* **2011**, *124*, 271–277. [CrossRef]

57. Chen, X.H.; Ma, L.H.; Dong, Y.W.; Song, H.; Pu, Y.; Zhou, Q.Y. Evaluation of the differences in phenolic compounds and antioxidant activities of five green asparagus (*Asparagus officinalis* L.) cultivars. *Qual. Assur. Saf. Crops Foods* **2017**, *9*, 479–487. [CrossRef]
58. Kobus-Cisowska, J.; Szymanowska, D.; Szczepaniak, O.M.; Gramza-Michałowska, A.; Kmiecik, D.; Kulczyński, B.; Szulc, P.; Górnaś, P. Composition of polyphenols of asparagus spears (*Asparagus officinalis*) and their antioxidant potential. *Ciência Rural* **2019**, *49*. [CrossRef]
59. Di Maro, A.; Pacifico, S.; Fiorentino, A.; Galasso, S.; Gallicchio, M.; Guida, V.; Severino, V.; Monaco, P.; Parente, A. Raviscanina wild asparagus (*Asparagus acutifolius* L.): A nutritionally valuable crop with antioxidant and antiproliferative properties. *Food Res. Int.* **2013**, *53*, 180–188. [CrossRef]
60. Ferrara, L.; Dosi, R.; Di Maro, A.; Guida, V.; Cefarelli, G.; Pacifico, S.; Mastellone, C.; Fiorentino, A.; Rosati, A.; Parente, A. Nutritional values, metabolic profile and radical scavenging capacities of wild asparagus (*A. acutifolius* L.). *J. Food Comp. Anal.* **2011**, *24*, 326–333. [CrossRef]
61. Adouni, K.; Júlio, A.; Santos-Buelga, C.; González-Paramás, A.M.; Filipe, P.; Rijo, P.; Lima, S.A.; Reis, S.; Fernandes, Â.; Ferreira, I.C.; et al. Roots and rhizomes of wild Asparagus: Nutritional composition, bioactivity and nanoencapsulation of the most potent extract. *Food Biosci.* **2022**, *45*, 101334. [CrossRef]
62. Dong, T.; Han, R.; Yu, J.; Zhu, M.; Zhang, Y.; Gong, Y.; Li, Z. Anthocyanins accumulation and molecular analysis of correlated genes by metabolome and transcriptome in green and purple asparaguses (*Asparagus officinalis* L.). *Food Chem.* **2019**, *271*, 18–28. [CrossRef]
63. Llorach, R.; Martínez-Sánchez, A.; Tomás-Barberán, F.A.; Gil, M.I.; Ferreres, F. Characterisation of polyphenols and antioxidant properties of five lettuce varieties and escarole. *Food Chem.* **2008**, *108*, 1028–1038. [CrossRef]
64. Medina-Lozano, I.; Bertolín, J.R.; Díaz, A. Nutritional value of commercial and traditional lettuce (*Lactuca sativa* L.) and wild relatives: Vitamin C and anthocyanin content. *Food Chem.* **2021**, *359*, 129864. [CrossRef]
65. Kim, D.-K. Antioxidative Components from the Aerial Parts of *Lactuca scariola* L. *Arch. Pharmacal Res.* **2001**, *24*, 427–430. [CrossRef] [PubMed]
66. Fan, L.; Zhang, S.; Yu, L.; Ma, L. Evaluation of antioxidant property and quality of breads containing Auricularia auricula polysaccharide flour. *Food Chem.* **2007**, *101*, 1158–1163. [CrossRef]
67. Kroh, L.W. Caramelisation in food and beverages. *Food Chem.* **1994**, *51*, 373–379. [CrossRef]
68. Delgado-Andrade, C.; Conde-Aguilera, J.A.; Haro, A.; de la Cueva, S.P.; Rufián-Henares, J.Á. A combined procedure to evaluate the global antioxidant response of bread. *J. Cereal Sci.* **2010**, *52*, 239–246. [CrossRef]
69. Mertens, D.R. Gravimetric determination of amylase-treated neutral detergent fiber in feeds with refluxing in beakers or crucibles: Collaborative study. *J. AOAC Int.* **2002**, *85*, 1217–1240.
70. Žilić, S. Phenolic Compounds of Wheat. Their Content, Antioxidant Capacity and Bioaccessibility. *MOJ Food Process. Technol.* **2016**, *2*, 00037. [CrossRef]
71. Abbas, S.R.; Sabir, S.M.; Ahmad, S.D.; Boligon, A.A.; Athayde, M.L. Phenolic profile, antioxidant potential and DNA damage protecting activity of sugarcane (*Saccharum officinarum*). *Food Chem.* **2014**, *147*, 10–16. [CrossRef] [PubMed]

metabolites

Article

Comparison of Black Tea Waste and Legume Roughages: Methane Mitigation and Rumen Fermentation Parameters

Gurkan Sezmis [1], Adem Kaya [2,*], Hatice Kaya [2], Muhlis Macit [2], Kadir Erten [3], Valiollah Palangi [4] and Maximilian Lackner [5,*]

1 Department of Animal Science, Faculty of Agriculture, Yozgat Bozok University, 66200 Yozgat, Türkiye
2 Department of Animal Science, Agricultural Faculty, Ataturk University, 25240 Erzurum, Türkiye
3 Department of Animal Science, Tekirdag Namik Kemal University, 59030 Tekirdag, Türkiye
4 Department of Animal Science, Faculty of Agriculture, Ege University, 35100 Izmir, Türkiye
5 Department of Industrial Engineering, University of Applied Sciences Technikum Wien, Hoechstaedtplatz 6, 1200 Vienna, Austria
* Correspondence: akaya@atauni.edu.tr (A.K.); maximilian.lackner@technikum-wien.at (M.L.)

Abstract: The chemical composition, in vitro total gas and CH_4 production and performance of cattle fed on factory black tea waste (*Camellia sinensis*) (BTW), alfalfa (*Medicago Sativa*), sainfoin (*Onobrychis sativa*) and white clover (*Trifolium repens*) was investigated. The gas production was quantified at the 24th hour of the incubation process. BTW was found to vary from roughages in chemical composition ($p < 0.05$). In addition, the roughages differed in terms of nutrient composition and gas production ($p < 0.05$). In legume roughages, acetic acid (AA), propionic acid (PA), butyric acid (BA), and total volatile fatty acids (TVFA) values ranged from 52.36–57.00 mmol/L, 13.46–17.20 mmol/L, 9.79–12.43 mmol/L, and 79.71–89.05 mmol/L, respectively. In comparison with black tea waste, legume roughages had higher values of AA, PA, BA, and TVFA. Black tea waste contained a higher acetic acid ratio than legume roughages when compared as a percentage. There was a similar ratio of propionic acid to the rate calculated for sainfoin (*Onobrychis sativa*) and clover (*Trifolium repens*), and a similar ratio of butyric acid to the ratio determined for alfalfa (*Medicago Sativa*). The current study shows that the 5.7–6.3% tannin content of black tea waste can be used in ruminant rations with high-quality roughages. Due to the fact that BTW reduces methane emissions from ruminants and eliminates energy waste from them, the environment can be improved. To obtain more reliable results, further animal feeding experiments on legume roughages and BTW are required.

Keywords: black tea waste; in vitro gas production; legume roughages; methane emission

1. Introduction

Ruminant animals need roughage due to the anatomical structure and physiological functions of their digestive systems. Roughage quality and the amount of roughage converted into yield are important factors for the animals. Those farms with insufficient or poor quality roughage sources require more concentrated feed sources for milk production, which increases ration costs and decreases profits. Consequently, dairy and beef cattle enterprises need affordable quality roughage sources. Several countries use alternative feed sources to supplement ruminant animals' nutrition to compensate for the lack of quality roughage.

It has also been proposed that BTW can be used to provide nutrients and bioactive compounds for the feeding of ruminants, which reduces waste and feed costs while maintaining a healthy environment [1,2]. Accordingly, the use of tea waste can be economically beneficial, and the tannin content should be considered [3]. Recently, one of the roughage sources that has attracted attention as an alternative has been factory black tea waste (BTW) [4,5]. After fresh tea is processed into products, factory black tea waste forms as a by-product, causing significant environmental pollution, including ecological and economic damage, since it is not used in any other way; It does not always end up as compost, but is also thrown into nature in an uncontrolled way [6], where methane can be formed. It is reported that this waste, which is a by-product of the food industry, contains 5.7–6.3% tannin, 2442 kcal ME/kg DM (ME = metabolic energy and DM = dry mass) energy, 19.8% crude fiber, and 18.2% crude protein (dry matter basis) [7]. Due to the content of tannins in black tea waste, a study conducted on black tea waste concluded that it reduces methane emissions and avoids ruminant energy consumption [8]. It is reported that 2–12% of the total energy taken with feed, and even 2–15% in some sources, is lost in the form of methane gas energy [9]. Feeds are fermented into gases such as volatile fatty acids (VFA), CO_2, and methane by rumen bacteria under rumen or in vitro rumen conditions. Rumen gas is produced primarily by the fermentation of carbohydrates into AA, PA, and butyric acids [10].

As part of ruminant nutrition, secondary plant compounds, such as tannins derived from tea leaves, can have a significant impact. Due to the chemical composition and concentration of tannins, there have been various studies with sometimes contradictory results regarding the performance of ruminants when tannins are co-fed. Among the many effects of tannins (high-molecular-weight polyphenols soluble in water) on ruminal fermentation, they decrease ruminal protein breakdown, reduce methane production, prevent bloating, and increase conjugated linoleic acid concentration in feeds derived from phenolic compounds [11,12]. Despite this, many studies have found that tannins may negatively affect the digestibility of diets. As tannins are interpreted differently depending on their type, origin, and supplement levels, their influence on ruminal protozoa, bacteria, fungi, and methanogens varies [13]. According to Petlum et al. [14], molecular weight greatly influences the effectiveness of condensed tannins as ruminal methane inhibitors. There has also been a correlation between tannin potency and dose. According to Mehansho et al. [15], hydrolyzable tannins are known to be toxic to ruminants as opposed to being digestion inhibitors.

Climate change has been exacerbated by greenhouse gas emissions produced by ruminants, including the greenhouse gases (GHGs) methane (CH_4), nitrous oxide, and carbon dioxide. The waste of energy caused by ruminant methane emissions is linked to global warming as well [16]. In addition to reducing feed energy loss and greenhouse gas emissions, animal nutritionists strive to maintain animal health, productivity, and performance. Therefore, ruminant production can be enhanced, as well as emissions of GHGs reduced, through reducing enteric CH_4 production. Hydrogen production efficiency in the rumen is different due to the end products of the microbial fermentation of nutrients. The rumen releases hydrogen through the production of acetate and butyrate, while propionate can be used as a competitive pathway. Ruminant methane mitigation strategies must consider a number of factors, including the metabolism of hydrogen and methanogens. The tannins found in most plants' skin, leaves, and roots are water-soluble phenolic compounds that form soluble or insoluble complexes with proteins. Tannins are chemically diverse groups of phenolic compounds. In addition to decreasing ruminal protein degradation and methanogenesis, tannins also prevent unsaturated fatty acids from being biohydrogenated in the rumen [17].

Using in vitro gas production (GP), the aim of this study was to evaluate BTW's potential use as a replacement for legume hays, such as alfalfa (*Medicago Sativa*), sainfoin (*Onobrychis sativa*), and clover (*Trifolium repens*). This can be achieved by analyzing and comparing in vitro total gas (GP) and methane (MG) productions, performance parameters, microbial protein production values (MP), and volatile fatty acids (VFA) of BTW and legume roughages.

2. Materials and Methods

2.1. Material

The alfalfa, sainfoin (1/10 flowering), and clover (general flowering) used in the present study were collected in Türkiye, from Atatürk University Plant Production Research and Application Center, and factory waste from black tea factories was also provided (Cumhuriyet, Gündoğdu, Pazarköy, Veliköy, and Camidağı) in the central district of Rize (at 39–56 north latitude and 32–51 east longitude) as a mixture of black tea waste that processes the fresh tea plants harvested in the second and third exile periods. To determine the parameters of in vitro GP, rumen liquor was collected from 7-year-old female Brown Swiss cattle, which completed its rumen development (as soon as the animals were humanely euthanized). In addition to being purebreds, each cow was registered with its respective breed association. Twice daily, at 8:00 a.m. and 04:00 p.m., cows were fed a mixture of rations, mostly corn silage. In order to maintain relatively stable rumen conditions, it was then placed in a screw-top glass bottle and transported to the lab in a capped thermos container containing water at about 39 °C. An anaerobic environment under CO_2 was used to filter rumen liquor for use in GP [16].

2.2. Method

In this study, alfalfa, sainfoin, white clover, and factory black tea wastes were dried at 65 °C, ground, and sieved at 1 mm with a 1 mm cutoff size. Dry and ground feed samples were chemically analyzed (DM, CP (976.05), EE (954.02), and crude ash (942.05) in accordance with AOAC (Association of Official Agricultural Chemists) guidelines [18]. NDF, ADF, and ADL are measured by the same methods described by Van Soest et al. [19] (NDF: neutral detergent fiber, ADF: acid detergent fiber, and ADL: acid detergent lignin).

In this study, fermentation parameters were determined from feed samples utilizing a technique described by Menke and Steingass [20]. Rumen fluid mixed with buffer solution (10 mL rumen fluid + 20 mL buffer solution) was prepared according to Menke et al. [21]. A milled feed sample (0.2 g) was incubated in 100 mL calibrated syringes containing rumen liquor for 3 replicates (3 for each incubation time with 3 blanks) to measure gas production. A correction factor of 49.61 mL per 0.200 g dry matter was applied to standard measurement conditions (University of Hohenheim). After reading the total amount of gas obtained for 24-h fermentation in the GP technique, accumulated gas was transferred to the infrared methane analyzer (Sensor Europa GmbH, Erkrath, Germany model), and the methane amount was determined as a percentage of the total gas [22] (Methane with a purity of 100% was used for calibration). Using Menke and Steingass' equations [20], performance parameters (ME, NE_L, and OMD) were calculated for feed samples.

$$\text{ME (MJ/kg DM)} = 2.20 + 0.136 \times \text{GP} + 0.057 \times \text{CP} + 0.0029 \times \text{CP}^2$$

$$\text{OMD (g/100 g DM)} = 14.88 + 0.889 \times \text{GP} + 0.45 \times \text{CP} + 0.0651 \times \text{XA}$$

$$\text{NE}_L \text{ (MJ/kg DM)} = 0.54 + 0.096 \times \text{GP} + 0.0038 \times \text{CP} + 0.000173 \times \text{CF}^2$$

where XA is ash (g/100 g DM), GP is the net gas production (mL) at 24 h, CP is crude protein, CF is crude fat, OMD is organic matter digestibility, ME is metabolic energy, and NE_L is net energy lactation. RFV index was calculated using the Rohweder et al. [23] formula after determining the DM intake, digestibility, and relative feed value (RFV) of samples.

DDMI or RFV ($g/kgW^{0.75}$) is a function of (DDM × DMI)/100.

DDM (%) is inversely related to ADF concentration, i.e., DDM decreases as ADF increases.

DMI ($g/kgW^{0.75}$) is inversely related to NDF concentration, i.e., DMI decreases as NDF increases.

The methods reported by Blümmel and Ørskov [24], Makkar et al. [25–27], and Van Soest and Robertson [28] were applied for the calculations of the amount of protein biomass from rumen bacteria, and the true digestible organic matter values of BTW and legume roughage were determined. In order to compute the TOMD values of the samples, the amount of sample weighed into the glass syringe was corrected according to OM [29].

$$\text{TOMD} = 100 - \left[\frac{D5 - D2}{D8 - D7} \times 100\right]$$

$D2$: bag weight after drying in an oven under 65 °C temperature; $D5$: ash weight after 4 h in an electric furnace with a temperature of 550 °C; $D8$: DM—Ash; and $D7$: DM of samples.

By gas chromatography, total volatile fatty acids (TVFA), acetic, propionic, and butyric acids in the rumen fluid remaining in the syringes were determined following the method by Wiedmeier et al. [30]. The following conditions were used for the GC-FID: initial temperature 50 °C, hold 2.0 min, and ramp rate 20 °C min^{-1}, and final temperature 280 °C, hold 0.5 min, and ramp rate 20 °C min^{-1}. At 240 °C for injection and 300 °C for the detector, the temperatures are similar. A split-less injection of 2 L was performed. In the run, N_2 was supplied at a rate of 2 mL min^{-1}. Flame ionization detectors were charged with nitrogen (30 milliliters per minute), hydrogen (44 milliliters per minute), and dry air (400 milliliters per minute).

An evaluation of tannins was completed by adding 6.25 mL of butanol-HCl reagent to 0.01 g of samples (95 mL butanol + 5 mL HCl + 0.05 g $Fe_2SO_4.7H_2O$). A boiling water bath (100 °C) was used for an hour; the tubes were removed, cooled, and centrifuged at 3000× g for 100 min. The absorbance at 550 nm was measured with a CE 2030 spectrophotometer (Cecil Instruments, United Kingdom) after the supernatant had been decanted into vials. A blank sample containing the reagent was included only in the measurements [25,26]. Microbial protein production values (MP) were calculated according to Makkar et al. [27]

2.3. Statistical Analysis

The SPSS 17.0 software package was used to analyze the data, and for the purpose of determining the difference between the means, this study used Duncan's multiple comparison test [31].

3. Results

3.1. Chemical Composition

Table 1 shows the average chemical composition of the feeds.

Table 1. Chemical compositions of feeds, (%).

Feed	DM	CA	CP	EE	NDF	ADF	ADL
BTW	92.19	4.59 ^{d}	16.51 ^{b}	1.35 ^{b}	60.21 ^{ab}	43.29 ^{a}	26.98 ^{a}
Sainfoin	92.05	7.24 ^{b}	15.97 ^{bc}	1.86 ^{b}	63.99 ^{a}	35.67 ^{b}	15.99 ^{b}
Clover	92.16	6.86 ^{c}	14.34 ^{c}	1.98 ^{ab}	58.60 ^{b}	32.57 ^{c}	14.27 ^{b}
Alfalfa	91.96	14.77 ^{a}	22.48 ^{a}	2.73 ^{a}	48.77 ^{c}	21.64 ^{d}	8.87 ^{c}
SEM	0.11	0.05	0.60	0.24	1.19	0.37	0.72
p	0.450	<0.001	<0.001	0.020	<0.001	<0.001	<0.001

a–d: means with different superscripts within same column are significantly different ($p < 0.05$). SEM: standard error of mean; DM: dry matter (%); CA: crude ash (%); CP: crude protein (%); EE: ether extract (%); NDF: neutral detergent fiber (%); ADF: acid detergent fiber (%); and ADL: acid detergent lignin (%).

Significant differences were found between BTW and legume roughage in terms of examined nutrients, except for dry matter content ($p < 0.05$). Legume roughages also showed differences among themselves with regard to chemical compounds ($p < 0.05$). In addition, the CT content of BTW and VFA (mg/L) contents were found to be 3.02 and 65.29, respectively.

3.2. In Vitro Gas (mL) and Methane Production (mL, %)

The amounts of gas and CH_4 produced by feeds are summarized in Table 2.

Table 2. Measurement values for gas and CH_4 production (mL/200 mg DM), methane ratios of feeds.

Feed	GP (mL)	Methane (mL)	Methane (%)
BTW	28.44 ^{c}	3.44 ^{c}	12.09 ^{b}
Sainfoin	59.04 ^{a}	8.46 ^{a}	14.34 ^{a}
Clover	54.00 ^{a}	8.03 ^{ab}	14.89 ^{a}
Alfalfa	42.84 ^{b}	6.88 ^{b}	15.95 ^{a}
SEM	2.15	0.44	0.52
p	<0.001	<0.001	0.005

a–c: means with different superscripts within the same column are significantly different ($p < 0.05$). SEM: standard error of mean; GP: gas production.

The measured differences among the legume hays and BTW, which had lower gas and CH_4 emissions values, were significant ($p < 0.01$). According to the ranking for gas and CH_4 emissions, sainfoin = clover > alfalfa > BTW. The observation that the BTW used in this study produced less gas than the legume roughages can be explained by the fact that it provides less useful carbohydrates for microorganisms.

3.3. MP, OMD, ME, and NEL Levels of Feeds

The values calculated for the MP and performance parameters of feeds are shown in Table 3 below.

Table 3. Means of MP and performance parameters of feeds.

Feeds	OMD (%)	ME (MJ/kg DM)	NE_L (MJ/kg DM)	MP (mg/mL)	IVTOMD (%)
BTW	40.71 ^{c}	7.01 ^{c}	3.87 ^{c}	108.49 ^{b}	46.87 ^{c}
Sainfoin	66.72 ^{a}	11.13 ^{a}	6.99 ^{a}	118.17 ^{b}	57.83 ^{b}
Clover	62.34 ^{a}	10.36 ^{a}	6.41 ^{ab}	154.03 ^{a}	54.65 ^{b}
Alfalfa	53.93 ^{b}	9.32 ^{b}	5.78 ^{b}	115.10 ^{b}	62.93 ^{a}
SEM	1.82	0.29	0.22	4.37	1.30
p	<0.001	<0.001	<0.001	<0.001	<0.001

a–c: means within column with unlike superscript differ significantly ($p < 0.05$). SEM: standard error of means; MP: microbial protein production; OMD: organic matter digestibility; ME: metabolic energy; NE_L: net energy lactation; and IVTOMD: in vitro true organic matter digestibility.

In terms of MP values, the feeds were arranged as clover > sainfoin = clover = BTW. Microbial protein production was higher in feeds with high energy and protein content. Elevated microbial protein production was detected in legume roughage plants, which have high energy and protein content and low cell wall components.

Concerning the parameters ME, OMD, and NEL, the ranking is sainfoin ≥ clover = alfalfa > black tea waste ($p < 0.01$). The reason for the high metabolic energy value in sainfoin and clover roughage is found in the amount of in vitro gas released as a result of fermentation. This is because the crude protein content was greater in these legumes, because the metabolic energy of the legume feeds was calculated by considering the 24-h gas production values, and crude protein contents were greater than those of BTW.

With respect to IVTOMD values (in vitro true organic matter digestibility), the order was determined as alfalfa > sainfoin = clover grass > black tea waste. The difference between BTW and legume roughage crops was significant ($p < 0.05$). Increases in the GP value at the 24th hour of fermentation and crude protein content of the feeds raised the IVTOMD.

3.4. DMD, DMI, and RFV Results of Feeds

In Table 4, calculated values for the DMD, DMI, and RFV parameters of the feeds are presented.

Table 4. DMD, DMI, and RFV values of feeds.

Feeds	DMD (%)	DMI	RFV (%)	FC
BTW	55.18 [d]	2.36 [a]	101.10 [bc]	3
Sainfoin	61.11 [c]	2.02 [b]	95.58 [c]	3
Clover	63.53 [b]	2.20 [ab]	108.22 [ab]	2
Alfalfa	72.04 [a]	2.12 [b]	118.61 [a]	2
SEM	0.29	0.06	3.29	
p	<0.001	0.024	0.006	

a–d: means within column with unlike superscript differ significantly ($p < 0.05$). SEM: standard error of means; DMD: dry matter digestibility; DMI: dry matter intake; RFV: relative feed value; and FC: forage classification.

Though RFV was developed to control the quality of Alfalfa in ABD, it is still in use for all roughages. Alfalfa is widely preferred by buyers because of its cost-effective production. Forage value is not only determined by RFV, but also by NDF and ADF based on alfalfa hay at full bloom quality. In order to calculate the RFV index, alfalfa hay is analyzed at full bloom for its forage quality.

3.5. The Effect of Feeds on Rumen Volatile Fatty Acid Levels

The average VFA levels formed by microbial fermentation in the rumen liquor in the GP method were determined and are shown in Table 5.

Table 5. Average values of volatile fatty acids composition of feeds.

Feed	AA mmol/L	PA mmol/L	BA mmol/L	AA %	PA %	BA %	TVFAmmol/lt
BTW	46.82 [c]	10.85 [d]	6.96 [c]	69.58 [a]	16.12 [b]	10.34 [b]	67.29 [c]
Sainfoin	52.36 [b]	13.46 [c]	10.37 [b]	65.71 [b]	16.87 [b]	13.01 [a]	79.71 [b]
Clover	57.00 [a]	15.06 [b]	12.43 [a]	65.23 [b]	17.24 [b]	14.22 [a]	87.38 [a]
Alfalfa	56.60 [a]	17.20 [a]	9.79 [b]	63.56 [c]	19.31 [a]	11.00 [b]	89.05 [a]
SEM	0.58	0.43	0.40	0.40	0.41	0.46	0.92
p	<0.001	<0.001	<0.001	<0.001	0.003	0.001	<0.001

a–d: means within column with unlike superscript differ significantly ($p < 0.05$). SEM: standard error of means; AA: acetic acid; PA: propionic acid; BA: butyric acid; TVFA: total volatile fatty acids.

The different values among the feeds were found to be significant regarding the amounts of AA, PA, BA, and TVFA ($p < 0.01$). Leguminous roughage had higher AA, PA, BA, and TVFA (mmol/L) than those of BTW. However, when compared as a percentage, the acetic acid ratio for black tea waste was found to be higher than the acetic acid ratio calculated for legume fodder plants. In addition, the ratio of propionic acid obtained for BTW was similar to the ratios assigned for sainfoin and clover.

4. Discussion

The crude protein value (16.51%) of factory black tea waste was between that determined for legume forages (14.34–22.48%). The CP values of legume roughage and BTW were in line with Filya et al. [32]. A roughage's protein content is a critical factor for microorganism growth [33]. Vegetable hay studied in this experiment, however, had a higher CP content than was required for optimal rumen microorganism activity. Significant differences existed between the feeds in terms of cell wall components ($p < 0.05$). Black tea waste's ADF and ADL contents were found to be higher than the values determined for legumes, while the NDF content was among the values determined for legumes. NDF values calculated for leguminous feeds were higher than those stated by Canbolat et al. [34]. The ADF and ADL content of legumes agreed with the findings of Filya et al. [31] and Canbolat et al. [34].

This can be explained by the fact that BTW is rich in ADF, ADL, and crude cellulose, from which microorganisms can benefit less. Moreover, GP and energy consumption are positively correlated [35]. In comparison to Filya et al. [32] and Canbolat et al. [34], BTW produced less gas and methane. A notable characteristic of tannins is that they block microbial enzyme activity due to their inhibitory enzyme activities [36]. Consequently, tannin compounds in tea waste may reduce the production of CH_4 during treatment, which is a desirable effect.

According to the present study, while high energy and protein content positively affected microbial protein production, increased cell wall components negatively influenced it. A similar result was obtained by Canbolat et al. [34].

Metabolic energy contents of legume-dried herbs were found to be in agreement with previous research findings [34,37]. The increase in the gas production value at the 24th hour of fermentation and the crude protein content of the feeds enhanced the OMD value. Additionally, NDF and ADF, which are hard to dissolve in the rumen, limit microbial fermentation, resulting in reduced OMD. OMD values obtained from the present study were in agreement with the values reported by Kamalak et al. [35].

Ruminant rations are restricted by low digestibility and voluntary intake, since ruminants' voluntary intake is influenced by the cell wall content and digestibility of roughages [34,38]. Alfalfa legume roughage had the lowest NDF, ADF, and ADL content and the highest IVTOMD ratio, while BTW showed the greatest cell wall components and lowest IVTOMD (in vitro true organic matter digestibility). Compared to previous studies, the present study found higher IVTOMD values [34,35,39]. It may be that the differences in roughages used as research materials in the studies explained the results obtained from the present study being greater than the values stated by the researchers.

As a result, an RFV of 100 is calculated for full-bloom alfalfa hay containing 41% ADF and 53% NDF [40]. Differences between feeds were found to be significant in relative feed values ($p < 0.01$). An increase in the components of the cell wall (NDF, ADF, and ADL), which makes digestion of feed difficult, adversely affects RFV. When RFV detected in legume roughage is compared to alfalfa in full bloom, it is seen that alfalfa and clover roughages are second quality, and sainfoin and BTW are third quality. The RFV value determined in legume roughage was found to be lower than that obtained by Adesogan et al. [41] and Canbolat et al. [34].

Compared to Palangi and Macit [16], the current study reported higher acetic acid ratios for legume feeds, and lower propionic acid and butyric acid ratios. In addition, while the acetic acid ratio calculated for black tea waste was similar to the findings reported by Palangi and Macit [16] for some legumes, the propionic acid and butyric acid ratios were found to be lower than in the reports of the researchers in question. A number of factors may have a bearing on the differences among the parameters mentioned in this work and those in previous studies, including animal, method, climate, soil structure, fertilization, plant spice and type, method of application, harvesting time, and vegetation period of feeds. Rumen fermentation and microbial activity can be indicated by changes in VFA profiles. The increase in acetate and reduction in propionate may only indicate an increase in hydrogen accumulation, not a change in methane production. During propionogenesis, hydrogen has not been fully incorporated because propionate has been reduced. That is the case because propiogenesis is a possible alternative receptor that can deliver hydrogen [42]. Consequently, since no increase in methane was observed in these treatments, it is possible that the secondary compounds in tea waste have restricted methanogen access to experimental feeds.

Based on Menke and Steingass [20], the gas produced is only affected by feed properties and chemical composition. According to the results of the present study, the constant ingredients of the tested feed, and the reduction in gas production by adding black tea waste extracts, it is likely that this reduction is a consequence of phenolic compounds' effect on ruminal microbial populations.

In ruminants, increased concentrations of condensed tannins stimulate protein passage through the rumen, which is probably due to reduced degradation rates by rumen microorganisms as well as decreased growth rates of proteolytic bacteria. Ruminant diets containing condensed tannins may also impair nutrient digestion [43,44] and fermentation [45]. It has been reported that livestock performance can be improved through the proper intake of tannin-containing compounds [46,47]. In light of this, it is critical to determine the appropriate dosage of this phenolic compound without adverse effects on digestion.

Tannins, one of the phenolic compounds found in BTW, can form hydrogen bonds with proteins, which are relatively stable within a pH range of 3.5–8, which prevents their ruminal degradation and reduces the rumen's ability to degrade nutrients. Once the feed reaches the abomasum, the pH drops below 3.5, breaking these bonds and making the compounds digestible. A tannin prevents bacteria from attaching to plant cell walls, which is crucial for the degradation process [48]. Additionally, complexes formed with proteins and carbohydrates restrict microorganisms' access to nutrients [49]. In addition to chelating agents, tannins also reduce the availability of metal ions, which rumen microorganisms need for metabolism [50]. Thus, it can be explained why tannin compounds in tea waste resulted in a reduction in methane production in the experiments.

5. Conclusions

It is desirable to utilize waste streams and aim for circularity; in this respect, the utilization of agro-industrial by-products to gain new feedstuff can reduce pollution, for instance, by avoiding uncontrolled disposal and CH_4 formation into the open atmosphere. Bacteria in fermentation require carbohydrates and protein in order to reproduce, so using these by-products helps preserve the environment and reduce production costs. The results obtained from the current work showed that black tea waste (BTW) has the potential to be used in ruminant rations together with quality roughage. Due to the tannin content of BTW, both the emission of methane gas and the energy loss in the form of methane gas from ruminants may be partially reduced by using it in ruminant rations. Black tea waste extracts could be used to increase feed use efficiency by limiting escaping nutrients from the rumen and preventing the ruminal breakdown of nutrients based on nutritional strategies. It is also necessary to conduct in vivo experiments to find out more about BTW. A large effect of tea waste addition was observed on gas volume, kinetic parameters, and methane production. An increase in methanotrophs is positively correlated with a reduction in methane production. Methane emissions can be reduced by microorganisms, especially those in the methanotroph group, but further research is needed to establish their role. This work has proven the potential of using black tea waste for economic and ecological benefits.

Author Contributions: Conceptualization, A.K.; methodology, G.S., K.E. and H.K.; software, V.P.; validation, A.K. and M.M.; formal analysis, A.K., H.K. and V.P.; investigation, M.M. and H.K.; resources, G.S., V.P. and K.E.; data curation, A.K.; writing—original draft preparation, A.K. and H.K.; writing—review and editing, M.L., M.M. and V.P.; supervision, M.M. All authors have read and agreed to the published version of the manuscript.

Funding: This research was supported by grant number FBA-2019-7388 from Atatürk University. Open Access Funding by the University of Applied Sciences Technikum Wien.

Institutional Review Board Statement: Not applicable.

Informed Consent Statement: Not applicable.

Data Availability Statement: No new data were created or analyzed in this study. Data sharing is not applicable to this article.

Acknowledgments: This experiment was supported by the Research Fund of Atatürk University with the FBA-2019-7388 project.

Conflicts of Interest: The authors declare no conflict of interest.

References

1. Salami, S.A.; Luciano, G.; O'Grady, M.N.; Biondi, L.; Newbold, C.J.; Kerry, J.P.; Priolo, A. Sustainability of feeding plant by-products: A review of the implications for ruminant meat production. *Anim. Feed Sci. Technol.* **2019**, *251*, 37–55. [CrossRef]
2. Georganas, A.; Giamouri, E.; Pappas, A.C.; Papadomichelakis, G.; Galliou, F.; Manios, T.; Zervas, G. Bioactive compounds in food waste: A review on the transformation of food waste to animal feed. *Foods* **2020**, *9*, 291. [CrossRef] [PubMed]
3. Kondo, M.; Hirano, Y.; Ikai, N.; Kita, K.; Jayanegara, A.; Yokota, H.O. Assessment of anti-nutritive activity of tannins in tea by-products based on in vitro rumen fermentation. *Asian-Australas. J. Anim. Sci.* **2014**, *27*, 1571–1576. [CrossRef] [PubMed]
4. Hassan, M.A.S.; Karslı, M.A. The Effects of Some Feed Additives in Nutrition of Ruminant Animals. *Int. J. Vet. Anim. Res.* **2022**, *5*, 107–112.
5. Wang, C.; Zhang, C.; Yan, T.; Chang, S.; Zhu, W.; Wanapat, M.; Hou, F. Increasing roughage quality by using alfalfa hay as a substitute for concentrate mitigates CH4 emissions and urinary N and ammonia excretion from dry ewes. *J. Anim. Physiol. Anim. Nut.* **2020**, *104*, 22–31. [CrossRef]
6. Abbood, H.N.; Hasnawi, S.F.; Hassain, A.S. The use of consumed black tea waste as natural adsorbent in removing the methylene blue dye. *Plant Arch.* **2020**, *20*, 2525–2529.
7. Çelebi, Ş.; Kaya, A.; Kaya, H.; Gül, M.; Yıldırım, B.A.; Macit, M. The effects of black tea factory waste supplementation into laying hen diets on performance, egg quality, yolk peroxidation, and blood parameters. *Kafkas Üni. Vet. Fak. Derg.* **2014**, *20*, 375–382. [CrossRef]
8. Nasehi, M.M.; Torbatinejad, N.M.; Rezaie, M.; Ghoorchi, T. Effect of polyethylene glycol addition on nutritive value of green and black tea co-products in ruminant nutrition. *Asian J. Anim. Vet. Adv.* **2017**, *12*, 254–260. [CrossRef]
9. Canbolat, Ö.; Kalkan, H.; Karaman, Ş.; Filya, İ. Esansiyel yağların sindirim, rumen fermantasyonu ve mikrobiyal protein üretimi üzerine etkileri. *Kafkas Üni. Vet. Fak. Derg.* **2011**, *17*, 557–565.
10. Besharati, M.; Palangi, V.; Salem, A.Z.M.; De Palo, P.; Lorenzo, J.M.; Maggiolino, A. Substitution of raw lucerne with raw citrus lemon by-product in silage: In vitro apparent digestibility and gas production. *Front. Vet. Sci.* **2022**, *9*, 1672. [CrossRef]
11. Koenig, K.M.; Beauchemin, K.A. Effect of feeding condensed tannins in high protein finishing diets containing corn distillers grains on ruminal fermentation, nutrient digestibility, and route of nitrogen excretion in beef cattle. *J. Anim. Sci.* **2018**, *96*, 4398–4413. [CrossRef] [PubMed]
12. Smeriglio, A.; Barreca, D.; Bellocco, E.; Trombetta, D. Proanthocyanidins and hydrolysable tannins: Occurrence, dietary intake and pharmacological effects. *Br. J. Pharmacol.* **2017**, *174*, 1244–1262. [CrossRef] [PubMed]
13. Gheshlagh, N.S.; Paya, H.; Taghizadeh, A.; Mohammadzadeh, H.; Palangi, V.; Mehmannavaz, Y. Comparative effects of extracted polyphenols from black and green tea wastes on in-vitro fermentability of feed ingredients. *Semin. Ciências Agrárias* **2021**, *42*, 2005–2022. [CrossRef]
14. Petlum, A.; Paengkoum, P.; Liang, J.B.; Vasupen, K. Molecular weight of condensed tannins of some tropical feed-leaves and their effect on in vitro gas and methane production. *Anim. Prod. Sci.* **2019**, *59*, 2154. [CrossRef]
15. Mehansho, H.; Butler, L.G.; Carlson, D.M. Dietary tannins and salivary proline-rich proteins: Interactions, induction, and defense mechanisms. *Annu. Rev. Nutr.* **1987**, *7*, 423–440. [CrossRef] [PubMed]
16. Palangi, V.; Macit, M. Indictable Mitigation of Methane Emission Using Some Organic Acids as Additives towards a Clean-er Ecosystem. *Waste Biomass Valori.* **2021**, *12*, 4825–4834. [CrossRef]
17. Denninger, T.; Schwarm, A.; Birkinshaw, A.; Terranova, M.; Dohme-Meier, F.; Münger, A.; Eggerschwiler, L.; Bapst, B.; Wegmann, S.; Clauss, M.; et al. Immediate effect of *Acacia mearnsii* tannins on methane emissions and milk fatty acid pro-files of dairy cows. *Anim. Feed Sci. Technol.* **2020**, *261*, 114388. [CrossRef]
18. AOAC. *Official Methods of Analysis of AOAC International*; AOAC International: Rockville, MD, USA, 2005.
19. Van Soest, P.J.; Robertson, J.B.; Lewis, B.A. Methods for dietary fiber, neutral detergent fiber and non-starch polysaccharides in ration to animal nutrition. *J. Dairy Sci.* **1991**, *74*, 3583–3597. [CrossRef]
20. Menke, K.H.; Steingass, H. Estimation of the energetic feed value obtained from chemical analysis and in vitro gas production using rumen fluid. *Anim. Res. Dev.* **1988**, *28*, 7–55.
21. Menke, K.H.; Raab, L.; Salewski, A.; Steingass, H.; Fritz, D.; Schneider, W. The estimation of the digestibility and metabolizable energy content of ruminant feedingstuffs from the gas production when they are incubated with rumen liquor in vitro. *J. Agric. Sci.* **1979**, *93*, 217–222. [CrossRef]
22. Goel, G.; Makkar, H.P.S.; Becker, K. Effect of *Sesbania sesban* and *Carduus pycnocephalus* leaves and Fenugreek (*Trigonella foenum-graecum* L.) seeds and their extract on partitioning of nutrients from roughage-and concentrate-based feeds to methane. *Anim. Feed Sci. Technol.* **2008**, *147*, 72–89. [CrossRef]
23. Rohweder, D.A.; Barnes, R.F.; Jorgensen, N. Proposed hay grading standards based on laboratory analyses for evaluating quality. *J Anim Sci.* **1978**, *47*, 747–759. [CrossRef]
24. Blümmel, M.; Ørskov, E.R. Comparison of in vitro gas production and nylon bag degradability of roughages in predicting of food intake in cattle. *Anim. Feed Sci. Technol.* **1993**, *40*, 109–119. [CrossRef]
25. Makkar, H.P.S.; Blummel, M.; Becker, K. Formation of complexes between polyvinyl pyrrolidone or polyethylene glycols and tannins and their implication in gas production and true digestibility in in vitro techniques. *Br. J. Nut.* **1995**, *73*, 897–913. [CrossRef]

26. Makkar, H.P.S.; Blümmel, M.; Becker, K. In vitro rumen apparent and true digestibilities of tannin-rich forages. *Anim. Feed Sci. Technol.* **1997**, *67*, 245–251. [CrossRef]
27. Makkar, H.P.S.; Sharma, O.P.; Dawra, R.K.; Negi, S.S. Simple determination of microbial protein in rumen liquor. *J. Dairy Sci.* **1982**, *65*, 2170–2173. [CrossRef] [PubMed]
28. Van Soest, P.J.; Robertson, J.B. *A Laboratory Manual for Animal Science 612*; Cornell University: Ithaca, NY, USA, 1985.
29. Canbolat, Ö. Yemlerin Bulundukları Yere Göre Örnek Alma ve Yem Analizleri. In *Yem Analiz Yöntemleri ve Yem Değerlendirme, 1*; Bakı Medyay Kitapevi: Bursa, Türkiye, 2019.
30. Wiedmeier, R.D.; Arambell, M.J.; Walters, J.L. Effect of orally administered pilocarpine on ruminal characteristics and nutrient digestibility in cattle. *J. Dairy Sci.* **1987**, *70*, 284–289. [CrossRef]
31. Snedecor, G.W.; Cochran, W. *Statistical Methods*; The Iowa State University Press: Ames, IA, USA, SPSS for windows release 17.0; SPSS Inc.: Chicago, IL, USA, 1976.
32. Filya, İ.; Karabulut, A.; Canbolat, Ö.; Değirmencioğlu, T.; Kalkan, H. Bursa bölgesinde yetiştirilen yem hammaddelerinin besleme değeri ve hayvansal organizmada optimum değerlendirilme koşullarının in vivo ve in vitro yöntemlerle saptanması üzerinde araştırmalar. *Uludağ Üni. Ziraat Fak. Bilim. Araşt. İncele. Ser.* **2002**, *25*, 1–16.
33. Blümmel, M.; Zerbini, E.; Reddy, B.V.S.; Hash, C.T.; Bidinger, F.; Khan, A.A. Improving the production and utilization of sorghum and pearl millet as livestock feed: Progress towards dual-purpose genotypes. *Field Crops Res.* **2003**, *84*, 143–158. [CrossRef]
34. Canbolat, Ö.; Kara, H.; Filya, İ. Bazı baklagil kaba yemlerinin in vitro gaz üretimi, metabolik enerji, organik madde sindirimi ve mikrobiyal protein üretimlerinin karşılaştırılması. *Uludağ Üni. Ziraat Fak. Derg.* **2013**, *27*, 71–81.
35. Kamalak, A.; Canbolat, Ö.; Gürbüz, Y.; Özay, O.; Erer, M.; Özkan, Ç.Ö. Kondense taninin rumimant hayvanlar üzerindeki etkileri hakkında bir inceleme. *KSÜ Fen Müh Derg.* **2005**, *8*, 132–137.
36. Palangi, V.; Taghizadeh, A.; Abachi, S.; Lackner, M. Strategies to mitigate enteric methane emissions in ruminants: A Review. *Sustainability* **2022**, *14*, 13229. [CrossRef]
37. Kamalak, A.; Canbolat, O.; Gürbüz, Y.; Erol, A.; Ozay, O. Variation in metabolizable energy content of forages estimated using in vitro gas production technique. *Pak. J. Bio. Sci.* **2004**, *7*, 601–605.
38. Karabulut, A.; Canbolat, O.; Kalkan, H.; Gurbuzol, F.; Sucu, E.; Filya, I. Comparison of in vitro gas production, metabolizable energy, organic matter digestibility and microbial protein production of some legume hays. *Asian-Australas. J. Anim. Sci.* **2007**, *20*, 517–522. [CrossRef]
39. Canbolat, Ö.; Karaman, Ş. Comparison of in vitro gas production, organic matter digestibility, relative feed value and metabolizable energy contents of some legume forages. *J. Agric. Sci.* **2009**, *15*, 188–195.
40. Redfearn, D.; Zhang, H.; Caddel, J.; Forage Quality Interpretations. Oklahoma Cooperative Extension Service F2117. 2017. Available online: http://pods.dasnr.okstate.edu/docushare/dsweb/Get/Document-2557/F-2117web.pdf.2006 (accessed on 30 August 2022).
41. Adesogan, A.T.; Sollenberger, L.E.; Moore, J.E. Forage Quality. In *Florida Forages Handbook*; Chambliss, C.G., Adjei, M.B., Eds.; University of Florida: Gainesville, FL, USA, 2006; Cooperative Extension Services.
42. Olijhoek, D.W.; Hellwing, A.L.F.; Brask, M.; Weisbjerg, M.R.; Højberg, O.; Larsen, M.K.; Dijkstra, J.; Erlandsen, E.J.; Lund, P. Effect of dietary nitrate level on enteric methane production, hydrogen emission, rumen fermentation, and nutrient digestibility in dairy cows. *J. Dairy Sci.* **2016**, *99*, 6191–6205. [CrossRef]
43. Derix, J. *The Effect of High Tannin Concentrations in Feed on Protein Digestion: Grazers Versus Intermediate Browsers*; Ghent University: Ghent, Belgium, 2017.
44. Samal, L.; Chaudhary, L.C.; Agarwal, N.; Kamra, D.N. Impact of phytogenic feed additives on growth performance, nutrient digestion and methanogenesis in growing buffaloes. *Anim. Prod. Sci.* **2018**, *58*, 1056–1063. [CrossRef]
45. Adejoro, F.A.; Hassen, A.; Akanmu, A.M. Effect of lipid-encapsulated acacia tannin extract on feed intake, nutrient digestibility and methane emission in sheep. *Animals* **2019**, *9*, 863. [CrossRef]
46. Zhang, M.; Zhang, Y.; Li, Z.; Zhang, C.; Tan, X.; Liu, X.; Lee, D.J. Anaerobic co-digestion of food waste/excess sludge: Substrates-products transformation and role of NADH as an indicator. *J. Environ. Man.* **2019**, *232*, 197–206. [CrossRef]
47. Costa, E.D.S.; Ribiero, C.V.D.M.; Silva, T.M.; Ribeiro, R.D.X.; Vieira, J.F.; Lima, A.D.O.; Oliveira, R.L. Intake, nutrient digestibility, nitrogen balance, serum metabolites and growth performance of lambs supplemented with *Acacia mearnsii* condensed tannin extract. *Anim. Feed Sci. Technol.* **2021**, *272*, 114744. [CrossRef]
48. Besharati, M.; Maggiolino, A.; Palangi, V.; Kaya, A.; Jabbar, M.; Eseceli, H.; Lorenzo, J.M. Tannin in Ruminant Nutrition. *Molecules* **2022**, *27*, 8273. [CrossRef] [PubMed]
49. Mueller-Harvey, I.; McAllan, A.B. Tannins: Their biochemistry and nutritional properties. *Adv. Plant Cell Biochem. Biotechnol.* **1992**, *1*, 151–217.
50. Scalbert, A. Antimicrobial properties of tannins. *Phytochemistry* **1991**, *30*, 3875–3883. [CrossRef]

metabolites MDPI

Article

Impacts of Different Prenatal Supplementation Strategies on the Plasma Metabolome of Bulls in the Rearing and Finishing Phase

Guilherme Henrique Gebim Polizel [1], Arícia Christofaro Fernandes [1], Édison Furlan [1], Barbara Carolina Teixeira Prati [1], José Bento Sterman Ferraz [2] and Miguel Henrique de Almeida Santana [1,*]

[1] Department of Animal Science, Faculty of Animal Science and Food Engineering—USP, Av. Duque de Caxias Norte, 225, Pirassununga 13635-900, SP, Brazil

[2] Department of Basic Sciences, Faculty of Animal Science and Food Engineering—USP, Av. Duque de Caxias Norte, 225, Pirassununga 13635-900, SP, Brazil

* Correspondence: mhasantana@usp.br; Tel.: +55-19-3565-4190

Abstract: This study investigated the effects of maternal nutrition on the plasma metabolome of Nellore bulls in the rearing and finishing phases, and metabolic differences between these phases. For this study, three nutritional approaches were used in 126 cows during pregnancy: NP—(control) mineral supplementation; PP—protein-energy supplementation in the final third; and FP—protein-energy supplementation during the entire pregnancy. We collected blood samples from male offspring in the rearing (450 ± 28 days old) and finishing phases (660 ± 28 days old). The blood was processed, and from plasma samples, we performed the targeted metabolome analysis (AbsoluteIDQ® p180 Kit). Multiple linear regression, principal component analysis (PCA), repeated measures analysis over time, and an enrichment analysis were performed. PCA showed an overlap of treatments and time clusters in the analyses. We identified significant metabolites among the treatments (rearing phase = six metabolites; finishing phase = three metabolites) and over time (21 metabolites). No significant metabolic pathways were found in the finishing phase, however, we found significant pathways in the rearing phase (Arginine biosynthesis and Histidine metabolism). Thus, prenatal nutrition impacted on plasma metabolome of bulls during the rearing and finishing phase and the different production stages showed an effect on the metabolic levels of bulls.

Keywords: beef cattle; fetal programming; metabolism; metabolomics; nutrigenomics

1. Introduction

The main economic interest traits in beef cattle production (growth and muscle development) are dependent on genetics and post-natal environmental conditions [1,2]. However, prenatal life can also impact these characteristics and others as immunology, stress response, reproduction, microbiota, organ development, and metabolic changes [3–12]. These alterations may persist long-term in an offspring's life—a concept known as fetal programming.

In beef cattle production, fetal programming may affect several phenotypes [13–16]. Nevertheless, the mechanisms that involve the responses to each prenatal environment, breed, and other variables have not been elucidated yet. Studies seeking molecular perspectives on phenotypes of interest in beef cattle, such as feed efficiency [17] and meat quality [18], have been increasing together with the decrease in the costs of omics technologies. The omics sciences refer to the study of physiological functions, biological processes, systems, and molecular structures [19]. Omics technologies primarily encompass genomics, transcriptomics, proteomics, and metabolomics [20,21]. The methods used by omics technologies (high-throughput technologies, next-generation sequencing, gas chromatography, liquid chromatography, mass spectrometry, etc.) are able to explore the genome, transcriptome, proteome, and metabolome more widely and efficiently, identifying biomarkers and making molecular diagnoses precisely [22].

The metabolome is the complete collection of metabolites present in a given tissue, cell, organ, or biofluid [23]. This omics science can investigate in depth the effects of prenatal nutrition on the progeny [24], and can significantly contribute to elucidating the metabolic mechanisms involved in the biological responses of the progeny to the nutritional stimulus. However, data from metabolomics are complex (systemic interactions, missing values, noise, etc.) and continuous improvements in pipelines are necessary to evaluate the information that will be generated. Through "machine learning" techniques (supervised and unsupervised), metabolomics data can have greater support for understanding the interactions of this dataset [25].

Thus, we hypothesized that different prenatal supplementation strategies influence the plasma metabolome of Nellore bulls in the rearing and finishing phase. Our objective was to evaluate the long-term effect of prenatal nutrition on plasma metabolome in each phase (rearing and finishing) and assess the metabolic differences over time through supervised and unsupervised machine learning techniques.

2. Material and Methods

2.1. Experimental Design

A herd of 126 Nellore cows was artificially inseminated using semen from four Nellore bulls. After pregnancy diagnosis, the cows were divided into three treatments (NP—Not programmed, PP—Partial Programming and FP—Full Programmed) based on age, body weight (BW) and body condition score. NP (control) cows received only mineral supplements throughout their pregnancy period (0.03% of BW per day). The PP treatment received protein-energy supplementation (0.3% of BW per day) only in the third trimester of pregnancy, while the FP group received this supplementation (0.3% of BW per day) from the confirmation of pregnancy (30 days) until calving. The three groups received mineral supplementation (0.03% of BW; already included in the protein-energy supplement; Table 1) for the entire period, a practice commonly performed in Brazil due to mineral deficiencies in tropical pastures, particularly during winter.

Table 1. Ingredients and nutrient content of the dams' supplement.

Ingredients	Mineral Supplement	Protein-Energy Supplement
Corn (%)	35	60
Soybean meal (%)	-	30
Dicalcium phosphate (%)	10	-
Urea 45% (%)	-	2.5
Salt (%)	30	5
Minerthal 160 MD (%) *	25	2.5
Total digestible nutrients (%)	26.76	67.55
Crude protein (%)	2.79	24.78
Non-protein nitrogen (%)	-	7.03
Acid detergent fiber (%)	1.25	4.76
Neutral detergent fiber (%)	4.29	11.24
Fat (%)	1.26	2.61
Calcium (g/kg)	74.11	6.2
Phosphorus (g/kg)	59.38	7.24

* Mineral premix composition (Minerthal company): Calcium = 8.6 g/kg; Cobalt = 6.4 mg/kg; Copper = 108 mg/kg; Sulfur = 2.4 g/kg; Fluorine = 64 mg/kg; Phosphorus = 6.4 g/kg; Iodine = 5.4 mg/kg; Manganese = 108 mg/kg; Selenium = 3.2 mg/kg; Zinc = 324 mg/kg; Sodium monensin = 160 mg/kg [10].

More information about the phenotypic effects of treatments on dams and details about the paddocks (*Brachiaria brizantha* cv. Marandu) can be found in Schalch Junior et al. [7].

After calving, dams and calves (both males and females) were kept together, despite the nutritional plan, until weaning at 240 ± 28 days old. During the calving to weaning period, all the cows received the same nutritional protocol (mineral supplementation of 0.03% of BW) as during the pregnancy period. After weaning, the calves were divided by sex (males and females), regardless of the treatment, and remained until the end of the rearing phase at 570 ± 28 days old. From calving, the male progeny was submitted to the same environmental conditions (sanitary and nutritional). During the rearing period, the young bulls received two types of supplements: an energetic supplement in the dry season (winter); and a protein supplement in the wet season (summer). Details of both supplements can be found in Polizel et al. [8]. From calving to the start of the finishing phase, the young bulls grazed on *Brachiaria brizantha* cv. Marandu pastures with water *ad libitum*.

The 63 bulls started the finishing phase (in feedlot paddocks) at 570 ± 28 days old and were slaughtered at 676 ± 28 days old by a pneumatic stunner. During this period, three distinct diets were given to the bulls: an adaptation diet (diet 1) for the first 15 days; a second one for the following 35 days; and a third one for the final 56 days. Once the finishing phase was complete, the animals were slaughtered at the FZEA/USP school slaughterhouse. The procedures (slaughter and processing of the carcasses) were performed according to the regulations established by the Ministry of Agriculture, Livestock and Supply of Brazil (MAPA, Normative Instruction No. 9 of 2004).

More details about the phenotypic effects on the bulls, and the finishing and rearing phase can be found in Polizel et al. [8–10].

2.2. Plasma Sample Collection and Processing

At 450 ± 28 days old and 660 ± 28 days old, the blood samples of the 63 bulls were collected. From this, 5 samples were randomly selected per treatment (n = 15) in the rearing phase, and another 5 samples from the same experimental units were selected in the finishing phase (n = 15) for carrying out this study. Blood was collected from the jugular vein in EDTA-coated tubes (BD Vacutainer, São Paulo, Brazil) and stored on ice until processing in the laboratory. The samples were centrifuged at 3000× g and 4 °C for 10 min within an hour of collection. The plasma supernatants were then transferred to fresh collection tubes and immediately frozen with dry ice before being stored at −80 °C until use.

2.3. Targeted Metabolomics

The AbsoluteIDQ® p180 Kit (Biocrates Life Sciences, Innsbruck, Austria) was employed for targeted metabolomics analysis of the plasma samples. This kit quantifies 188 metabolites, including 21 amino acids, 21 biogenic amines, 40 acylcarnitines (Cx:y), 14 lysophosphatidylcholines (lysoPC), 76 phosphatidylcholines (PC), and 15 sphingolipids (SMx:y). The analysis was conducted by Apex Science (Campinas, São Paulo, Brazil). The kit is a combined flow injection (FIA) and liquid chromatography (LC) tandem mass spectrometry assay. The amino acids and biogenic amines were analyzed by liquid chromatography tandem-mass spectrometry (HPLC-MS/MS) with electrospray ionization. The lysophosphatidylcholines, phosphatidylcholines, acylcarnitines, and hexose were evaluated by flow injection analysis-tandem mass spectrometry (FIA-MS/MS). Internal standards, analyte derivatization, and metabolite extraction are integrated into a 96-well plate kit. Mass detection and compound identification were performed by multiple reaction monitoring. Briefly, after the addition of 10 µL of supplied internal standard solution to each well on the filter spot of the 96-well extraction plate, 10 µL of each plasma sample, quality control (QC) samples, or calibration standard was added to the appropriate wells. The plate was dried under a gentle stream of nitrogen. Then, amino acid and biogenic amines were derivatized with phenyl isothiocyanate (Sigma Aldrich, Germany), and dried again. Metabolite extraction was performed with 5 mM ammonium acetate in methanol. The final extracts were analyzed after appropriate dilution by HPLC-MS/MS (amino acids and biogenic amines) and FIA-MS/MS (lysophosphatidylcholines, phosphatidylcholines, acylcarnitines, and hexose). The software MetIDQ® v1.0 performed the metabolite quantification and quality assessment. Biocrates experimentally determines the metabolite-specific limits of detection (LOD) of the assay.

2.4. Statistical Analysis

Data processing and the univariate analysis (supervised technique; multiple linear regression) were performed using the "LM" function in the R software environment (version 4.1.2) (https://www.r-project.org/ accessed on 1 January 2023). Metabolites with more than 70% of samples below LOD were removed (filtering data) from the dataset (Rearing phase = 168 metabolites remaining; and Finishing phase = 171 metabolites remaining). The LOD values that remained in the dataset after filtering were replaced by the mean of each variable.

The statistical model used in both production phases (rearing and finishing phase) was:

$$Y_{jk} = \mu + \beta_1 Age_{b1} + Treat_j + e_{jk} \tag{1}$$

where: Y_{jk} are the observed metabolite from k^{th} animal, recorded on j^{th} treatment; μ is a constant; β_1 is the regression coefficient of covariate animal's age; Age_{b1} is the observed value for bull's age of kth animal; $Treat_j$ is the fixed effect of jth treatment; and e_{jk} is the residual random term. The residuals were tested for homoscedasticity (Levene's test) and for normality (Shapiro-Wilk test), and the differences between treatments were considered significant when $p \leq 0.05$ by the Tukey Kramer test.

In addition, the metabolite concentration table was uploaded to MetaboAnalyst 5.0 [26], and the data were Auto-scaled (mean-centered and divided by the standard deviation of each variable) before analysis. We performed a principal component analysis (PCA; unsupervised method) of each phase and between both (rearing and finishing phase), an enrichment analysis and a repeated measures analysis over time. The PCA was performed to assess the clusters between treatments (NP, PP, and FP) in each production phase and to evaluate the differences between these stages. The enrichment analysis was carried out by MetaboAnalyst to identify the most relevant biological processes associated with the differentially expressed metabolites (identified in univariate analysis) based on the Kyoto Encyclopedia of Genes and Genomes database (KEGG Pathway). Biological processes with $p \leq 0.05$ were considered significant. The repeated measures analysis over time was performed by the linear model with the covariate adjustment function of MetaboAnalyst. This analysis considered variables in the model just the time, treatment, and the interaction between both. This approach allows using linear models (limma or lm) to perform significance testing with covariate adjustment.

3. Results

3.1. Unsupervised Analysis of Metabolome (PCA)

The results found in PCA of the plasma metabolome in the rearing phase (Figure 1), finishing phase (Figure 2), and between these both phases (Figure 3) are similar. The distribution of all analyses data showed an overlap between all groups and it was not possible to observe a clustering among the treatments or between the different production stages. This may indicate that the metabolite profile presented only a few or no variables expressed differentially among treatments. In the rearing phase, the two principal components together explain 50.6% of the total variance (PC1 = 32.6%; PC2 = 18.0%). In the finishing phase, the two principal components together explain 51.4% of the total variance (PC1 = 31.3%; PC2 = 20.3%). Between the times (rearing and finishing phases), the two principal components together explain 46.1% of the total variance (PC1 = 30.3%; PC2 = 15.8%).

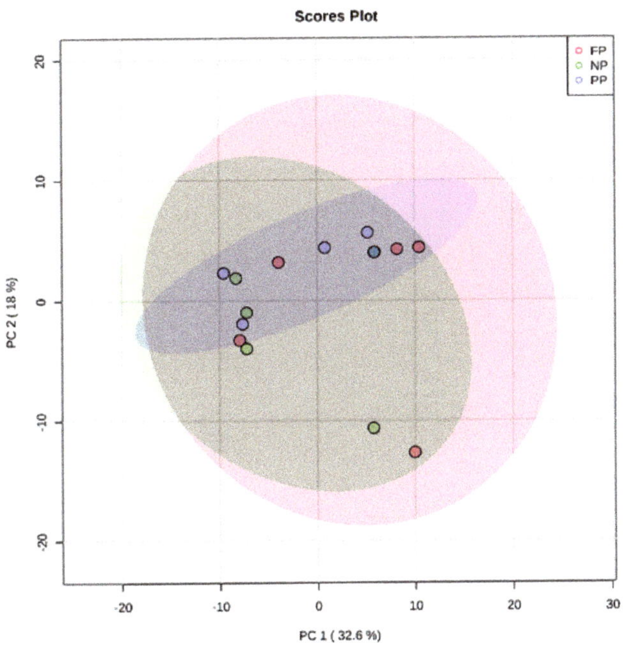

Figure 1. Principal component analysis (PCA) scores plot of metabolome distribution of bulls' blood plasma in the rearing phase among the treatments (NP, PP, and FP).

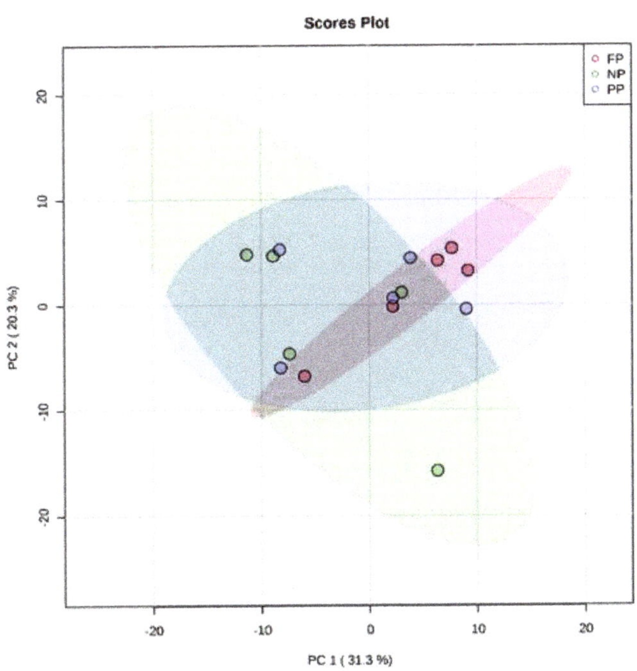

Figure 2. Principal component analysis (PCA) scores plot of metabolome distribution of bulls' blood plasma in the finishing phase among the treatments (NP, PP and FP).

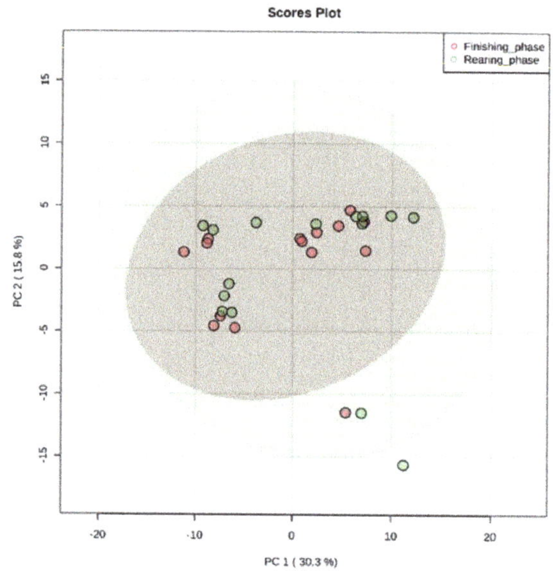

Figure 3. Principal component analysis (PCA) scores plot of metabolome distribution of bulls' blood plasma between the different production stages (finishing phase and rearing phase).

3.2. Supervised Analysis of Metabolome (Multiple Linear Regression)

Regarding the results found in the supervised analysis of the rearing phase (Table 2), six metabolites were identified as differentially expressed according to the prenatal treatment received. The NP group had higher levels of plasmatic carnosine than the groups that received protein-energy supplementation (PP and FP). The NP treatment also showed higher levels of putrescine than the PP group and higher levels of Trans-4-Hydroxy-L-Proline (t4-OH-Pro) and tryptophan than the FP treatment. Citrulline, on the other hand, showed higher levels in the FP treatment and showed a significant difference with the NP treatment. Finally, SM C18:1 showed a significant difference between FP and PP treatments, with the FP treatment having the highest concentrations.

Table 2. Plasma significant metabolites (μM; mean ± standard error) of bulls in the rearing phase submitted to the different prenatal nutrition approaches (NP, PP, and FP) with their respective *p* values.

Metabolites	NP	PP	FP	*p* Value
Carnosine	23.69 ± 0.749 [a]	18.68 ± 0.882 [b]	19.44 ± 1.111 [b]	0.008
Putrescine	0.128 ± 0.014 [a]	0.076 ± 0.011 [b]	0.090 ± 0.005 [ab]	0.020
t4-OH-Pro	41.66 ± 2.435 [a]	35.76 ± 1.985 [ab]	33.76 ± 0.838 [b]	0.032
Tryptophan	41.57 ± 3.867 [a]	38.70 ± 2.632 [ab]	31.26 ± 1.178 [b]	0.037
Citrulline	48.51 ± 2.712 [a]	52.72 ± 3.223 [ab]	59.54 ± 1.657 [b]	0.041
SM C18:1	6.805 ± 2.072 [ab]	1.817 ± 0.997 [a]	9.108 ± 2.147 [b]	0.047

The small letters overwritten represent the significant contrasts. NP—not programmed; PP—partial programming; FP—fully programmed.

In the finishing phase (Table 3), just three metabolites were differentially expressed among the treatments. C5:1-DC was the most significant metabolite among the treatment ($p = 0.001$), where the group FP showed higher levels compared to other treatments (NP and PP). The FP treatment also had higher levels of SM C26:0 compared to NP and PP groups. Lastly, the group PP showed higher levels of Serotonin in comparison to the group not supplemented (NP), and showed no differences with FP treatment.

Table 3. Plasma significant metabolites (μM; mean ± standard error) of bulls in the rearing phase submitted to the different prenatal nutrition approaches (NP, PP, and FP) with their respective *p* values.

Metabolites	NP	PP	FP	*p* Value
C5:1-DC	0.008 ± 0.006 [a]	0.005 ± 0.005 [a]	0.012 ± 0.009 [b]	0.001
SM C26:0	0.068 ± 0.032 [a]	0.049 ± 0.006 [a]	0.384 ± 0.090 [b]	0.002
Serotonin	0.414 ± 0.047 [a]	0.910 ± 0.176 [b]	0.646 ± 0.080 [ab]	0.026

The small letters overwritten represent the significant contrasts. NP—not programmed; PP—partial programming; FP—fully programmed.

The results of all expressed metabolites in both production stages can be better visualized in Table S1 (rearing phase) and Table S2 (finishing phase).

3.3. Repeated Measures Analysis over Time

Regarding the results in the repeated measures analysis over time, we found 21 significant metabolites for the variable "Time" and none for the interaction of "Treatment × Time" (all p values > 0.05). The significant results can be viewed in Table 4 and all the results regardless of the p value can be found in Table S3.

Table 4. Repeated measures analysis over time with the significant metabolites at least in one *p* value.

Metabolites	*p* Values	
	Time	Interaction (Treatment × Time)
Taurine	<0.001	0.593
Carnosine	<0.001	0.933
Histidine	<0.001	0.825
Proline	<0.001	0.626
Sarcosine	<0.001	0.969
Glutamine	<0.001	0.406
Tryptophan	<0.001	0.544
Serine	<0.001	0.825
Leucine	<0.001	0.368
Ornithine	0.001	0.145
Asparagine	0.001	0.968
Met-SO	0.001	0.175
Arginine	0.001	0.459
Creatinine	0.002	0.184
Lysine	0.002	0.474
Methionine	0.004	0.781
ADMA	0.009	0.963
Glycine	0.009	0.075
Kynurenine	0.015	0.268
Citrulline	0.023	0.071
C9	0.026	0.820

3.4. Functional Enrichment

In the enrichment analysis of plasma metabolites of bulls, we found the top biological processes related to differentially expressed metabolites among the prenatal treatments in the rearing phase (Figure 4) and in the finishing phase. In the finishing phase, as we found just three differential metabolites expressed, it was identified no significant biological processes related. However, in the rearing phase we found two significant processes related to the set of differentially expressed metabolites. Between them, Arginine biosynthesis ($p = 0.036$) and Histidine metabolism ($p = 0.041$) were considered the 2 enriched significant metabolic processes.

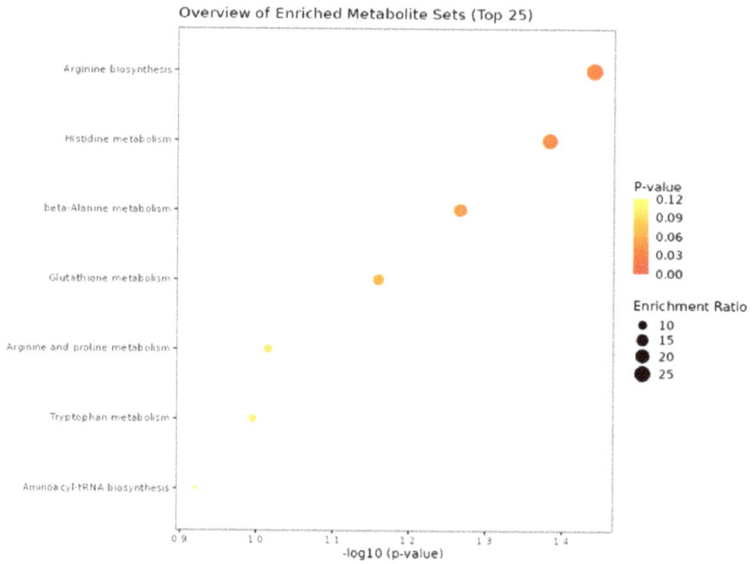

Figure 4. Top biological processes involved with significant blood plasma metabolites of bulls in the rearing phase from the three different maternal treatments (NP, PP, and FP).

4. Discussion

According to our literature search, this is the first study that assessed the impact of three different prenatal supplementation approaches on the plasma metabolome of Nellore bulls in the rearing and finishing phase. This study is innovative and may contribute to the understanding of some molecular mechanisms involving maternal nutrition and long-term effects on offspring in beef cattle.

Based on our results, we selected the main classes and metabolites differentially expressed among the treatments to discuss briefly. In addition, we also discuss the metabolic pathways and the differences found between the production phases (rearing and finishing).

Carnosine is a dipeptide composed of beta-alanine and histidine. In cattle, this metabolite is found in greater amounts in skeletal muscle [27] and has several functions related to homeostasis and epigenetic regulation [28–31]. Furthermore, carnosine levels have already been positively correlated with feed efficiency in Nellore cattle [32]. Thus, observing only the levels of this metabolite, we could conclude that animals from treatments that received protein-energy supplementation during the prenatal period (PP and FP) may have lower feed efficiency than animals from the control treatment (NP) in the rearing phase. However, complex phenotypes, such as feed efficiency, are controlled by several genetic mechanisms, which makes it difficult to conclude based only on the concentration level of one metabolite.

Putrescine is a polyamine derived from Arginine, naturally found in all organisms [33]. This metabolite has some functions related to epigenetic mechanisms (DNA topology; [34]) and can improve the response of proteins to heat shock [35]. Additionally, according to Liao et al. [36], putrescine levels may also be related to adaptability to heat stress in cattle. Thus, the results of our study may be indicative that progenies from cows that did not receive protein-energy supplementation (NP) may have lower susceptibility to heat stress than the progenies of the PP group (supplementation in the final third of pregnancy) in the rearing phase.

T4-OH-Pro is a biogenic amine that is associated with type 2 diabetes mellitus [37] and with an increased risk of prostate cancer [38] in humans. In pregnant cattle, T4-OH-Pro in blood serum was associated with the group of cows with higher body condition scores in early lactation [39]. In addition, a correlation between the blood concentration of T4-

OH-Pro and the level of lipolysis in early lactation in cows has also been reported [40]. In our study, the control treatment (NP) during the rearing phase showed the highest levels of this metabolite and differed from the treatment in which it received energy protein supplementation throughout pregnancy (FP). This may be indicative that progenies from FP cows may have lower rates of lipolysis than the NP group.

Citrulline is a non-essential amino acid considered a biomarker for enterocyte mass, epithelial cell damage, and absorptive function [41,42]. The administration of citrulline to pregnant ewes and cows improves embryo survival [43] and tends to increase the number of antral follicles in the ovaries of ewes [44]. Furthermore, citrulline has several functions: protein synthesis, intestinal homeostasis, nitrogen balance, growth and development, anti-oxidation, muscle performance, intestinal functions, renal function, exercise performance, blood pressure, vasodilation, and anti-inflammatory action [45–48]. Given these numerous functions, in the rearing phase, animals from the FP treatment (higher concentration of citrulline) may have advantages over the control treatment (NP). This is mainly related to growth and muscle performance, which are the central objectives of beef cattle production.

Tryptophan is an essential amino acid that plays a major role in protein synthesis [49]. This metabolite also has several other important functions and metabolic pathways related to serotonin synthesis [50,51], kynurenine synthesis [52,53], and synthesis of melatonin [49]. These biological pathways are related to immunological modulatory effects [54]. The higher level of tryptophan found in the NP treatment animals may indicate greater resilience to stressors compared to the FP treatment in the rearing phase.

Serotonin, or 5-Hydroxytryptamine, is a biogenic amine with several effects on the central nervous system [55], and functions in the regulation of energy metabolism, lactation, and calcium homeostasis [56,57]. The synthesis of serotonin occurs naturally using tryptophan in the diet [50]. Specifically, in animal production, the roles of serotonin are related to feed intake control [58], energy metabolism [59], stress [60], immunological system [61], mineral homeostasis [62], and hormone release [63]. Serotonin has been studied in cattle, but the metabolism and manipulation of this metabolite have received limited investigations [64].

C5:1-DC (Glutaconylcarnitine) is a metabolite belonging to the acylcarnitine class. Acylcarnitines are esters of fatty acids and L-carnitine [65]. Due to a large number of constituents and structures, the acylcarnitines play important roles in cell metabolism [66]. Among its specific functions have: regulating the balance of intracellular sugar and lipid metabolism [67], metabolism of branched-chain amino acids [68], homeostasis of the mitochondrial acyl-CoA/CoA ratio, regulation of glucagon/insulin [69], oxidation of fatty acids [70], and others. This metabolite class is related to several metabolic diseases [71], however, the relationship with animal husbandry is scarce in the literature. According to Ladeira et al. and Nguyen et al. [72,73], acylcarnitine is associated with the deposition of intramuscular fat in beef cattle, which may affect the marbling of the meat. In our study, the different levels of C5:1-DC found among FP treatment and the others (PP and NP) may be related mainly with marbling and with the metabolism of amino acids, as we identified in the significant biological processes in the rearing phase.

SM C18:1 and SM C26:0 are types of sphingolipids of the sphingomyelin class. Among the sphingolipids, this class of metabolites is the most abundant component of the cellular plasma membrane in mammals, preferentially associating with cholesterol to form stronger lipid bonds [74]. These bonds play a major role in signal transduction in many cell types and in maintaining membrane integrity [75–77]. Sphingomyelins are also correlated with parameters of obesity, insulin resistance, liver function, and lipid metabolism [78]. More studies are still needed in the area of prenatal nutrition and animal production in order to understand the maternal effects on the offspring's metabolically levels of sphingomyelins. Additionally, the other classes of metabolites discussed here in this study need deeper information about their functions and roles in animal production and fetal programming. With this, it will be possible, in fact, to understand the molecular mechanisms that involve the phenotype.

Regarding the metabolome differences found between the production stages (rearing and finishing phases), we already expected plasmatic concentration levels to divergent in some metabolites. The different environments which the bulls were exposed (extensive production x intensive production) when comparing the period of the rearing phase and finishing phase were completely different in terms of nutrition and stressful factors. Due to the use of a greater proportion of grains and other products during the finishing phase, these foods can lead to digestive disorders in animals [79]. Other critical points are related to high farm animal density in the feedlot and the process of adaptation in a new environment [80], which may affect the productive indices of the herd and trigger stressful responses. All of these factors impact the metabolic state of the animals, showing differences in the plasma metabolome when we compare the rearing phase and finishing phase. Furthermore, we have not found an interaction between time and treatment in the analysis. Thus, the effect that we observed in the repeated measures analysis over time is just related to the difference caused by the production stages.

In the enrichment analysis, the significant biological processes affected by prenatal treatments in the rearing phase were related to Arginine biosynthesis and Histidine metabolism. No significant metabolic pathway was found in the finishing phase due to the lower number of differential metabolites expressed among the treatments (only 3 metabolites). The significant metabolic pathways identified are related to amino acid metabolism, therefore implying changes in protein metabolism. As the different maternal supplementation approaches (NP, PP, and FP) were based on protein and energy levels, this impacted on the protein metabolism in the rearing phase (Arginine biosynthesis and Histidine metabolism). According to Schalch Junior et al. [7], the top metabolic pathway affected by prenatal nutrition in calves (30 days old) was the Histidine metabolism, showing similar effects with the present study when compared to the rearing phase.

5. Conclusions

The different prenatal supplementation approaches (NP, PP, and FP) influenced the plasma metabolome of bulls in both the rearing phase and the finishing phase. In addition, we found biological pathways (amino acid metabolism) affected by prenatal nutrition in the rearing phase and metabolites differentially expressed over time (between the rearing phase and finishing phase). The most significant metabolites found in the study are related to protein metabolism, which could be expected due to the protein supply differences in the prenatal nutritional approaches. In summary, these findings corroborate the understanding of the part of the molecular mechanisms that involve fetal programming.

Supplementary Materials: The following supporting information can be downloaded at: https://www.mdpi.com/article/10.3390/metabo13020259/s1, Table S1: Plasma metabolites (μM; mean ± standard error) of bulls in the rearing phase submitted to the different prenatal nutrition approaches (NP, PP and FP) with their respective *p* values. Table S2: Plasma metabolites (μM; mean ± standard error) of bulls in the rearing phase submitted to the different prenatal nutrition approaches (NP, PP and FP) with their respective *p* values. Table S3: Repeated measures analysis over time with all metabolites and the respective *p* values.

Author Contributions: Conceptualization, M.H.d.A.S.; methodology, G.H.G.P. and J.B.S.F.; formal analysis, G.H.G.P.; investigation, G.H.G.P., A.C.F., É.F. and B.C.T.P.; writing—original draft preparation, G.H.G.P.; writing—review and editing, M.H.d.A.S. and J.B.S.F.; supervision, M.H.d.A.S.; project administration, M.H.d.A.S.; funding acquisition, M.H.d.A.S. and G.H.G.P. All authors have read and agreed to the published version of the manuscript.

Funding: This research was funded by São Paulo Research Foundation (FAPESP), grant number (2017/12105-2 and 2021/03265-1) and National Council for Scientific and Technological Development (CNPq), grant number 307593/2021-5.

Institutional Review Board Statement: The Research Ethics Committee of Faculty of Animal Science and Food Engineering from University of São Paulo approved this study, under protocol No. 1843241117, according to the guidelines of the National Council for the Control of Animal Experimentation.

Informed Consent Statement: Not applicable.

Data Availability Statement: The data presented in this study are available on request from the corresponding author. The data are not publicly available due to privacy.

Acknowledgments: The authors acknowledge the São Paulo Research Foundation (FAPESP, National Council for Scientific and Technological Development (CNPq) (grant number 307593/2021-5 and the College of Animal Science and Food Engineering (FZEA-USP)].

Conflicts of Interest: The authors declare no conflict of interest.

References

1. Warner, R.D.; Greenwood, P.L.; Pethick, D.W.; Ferguson, D.M. Genetic and Environmental Effects on Meat Quality. *Meat Sci.* **2010**, *86*, 171–183. [CrossRef] [PubMed]
2. Connolly, S.M.; Cromie, A.R.; Berry, D.P. Genetic Differences Based on a Beef Terminal Index Are Reflected in Future Phenotypic Performance Differences in Commercial Beef Cattle. *Animal* **2016**, *10*, 736–745. [CrossRef]
3. Clavé, A.; Ripoll, G.; Casasús, I.; Sanz, A. Long-Term Effects of Early Maternal Undernutrition on the Growth, Physiological Profiles, Carcass and Meat Quality of Male Beef Offspring. *Res. Vet. Sci.* **2022**, *142*, 1–11. [CrossRef]
4. Long, N.M.; Prado-Cooper, M.J.; Krehbiel, C.R.; Desilva, U.; Wettemann, R.P. Effects of Nutrient Restriction of Bovine Dams during Early Gestation on Postnatal Growth, Carcass and Organ Characteristics, and Gene Expression in Adipose Tissue and Muscle. *J. Anim. Sci.* **2010**, *88*, 3251–3261. [CrossRef] [PubMed]
5. Coloma-García, W.; Mehaba, N.; Llonch, P.; Caja, G.; Such, X.; Salama, A.A.K. Prenatal Heat Stress Effects on Gestation and Postnatal Behavior in Kid Goats. *PLoS ONE* **2020**, *15*, e0220221. [CrossRef] [PubMed]
6. Cavalcante, C.R.; de Oliveira, B.F.; Gebim, P.G.H.; Édison, F.; Pontes, G.N.; Schmidt, P.D.A.; Guilherme, P.; de Almeida, S.M.H. Effects of Maternal Nutrition on Female Offspring Weight Gain and Sexual Development. *Front. Genet.* **2021**, *12*, 2059. [CrossRef]
7. Schalch, F.J., Jr.; Polizel, G.H.G.; Cançado, F.A.C.Q.; Fernandes, A.C.; Mortari, I.; Pires, P.R.L.; Fukumasu, H.; Santana, M.H.d.A.; Saran Netto, A. Prenatal Supplementation in Beef Cattle and Its Effects on Plasma Metabolome of Dams and Calves. *Metabolites* **2022**, *12*, 347. [CrossRef]
8. Polizel, G.H.G.; Cançado, F.A.C.Q.; Dias, E.F.F.; Fernandes, A.C.; Cracco, R.C.; Carmona, B.T.; Castellar, H.H.; Poleti, M.D.; de Almeida Santana, M.H. Effects of Different Prenatal Nutrition Strategies on the Liver Metabolome of Bulls and Its Correlation with Body and Liver Weight. *Metabolites* **2022**, *12*, 441. [CrossRef]
9. Polizel, G.H.G.; Espigolan, R.; Fantinato-Neto, P.; de Francisco Strefezzi, R.; Rangel, R.B.; de Carli, C.; Fernandes, A.C.; Dias, E.F.F.; Cracco, R.C.; de Almeida Santana, M.H. Different Prenatal Supplementation Strategies and Its Impacts on Reproductive and Nutrigenetics Assessments of Bulls in Finishing Phase. *Vet. Res. Commun.* **2022**, 1–15. [CrossRef]
10. Polizel, G.H.G.; de Francisco Strefezzi, R.; Cracco, R.C.; Fernandes, A.C.; Zuca, C.B.; Castellar, H.H.; Baldin, G.C.; de Almeida Santana, M.H. Effects of Different Maternal Nutrition Approaches on Weight Gain and on Adipose and Muscle Tissue Development of Young Bulls in the Rearing Phase. *Trop. Anim. Health Prod.* **2021**, *53*, 1–9. [CrossRef]
11. Polizel, G.H.G.; Fantinato-Neto, P.; Rangel, R.B.; Grigoletto, L.; de Oliveira Bussiman, F.; Cracco, R.C.; Garcia, N.P.; Ruy, I.M.; Ferraz, J.B.S.; de Almeida Santana, M.H. Evaluation of Reproductive Traits and the Effect of Nutrigenetics on Bulls Submitted to Fetal Programming. *Livest. Sci.* **2021**, *247*, 104487. [CrossRef]
12. Elolimy, A.; Alharthi, A.; Zeineldin, M.; Parys, C.; Helmbrecht, A.; Loor, J.J. Supply of Methionine During Late-Pregnancy Alters Fecal Microbiota and Metabolome in Neonatal Dairy Calves Without Changes in Daily Feed Intake. *Front. Microbiol.* **2019**, *10*, 2159. [CrossRef]
13. Block, J.J.; Webb, M.J.; Underwood, K.R.; Gonda, M.G.; Harty, A.A.; Salverson, R.R.; Funston, R.N.; Olson, K.C.; Blair, A.D. Influence of Maternal Protein Restriction in Primiparous Beef Heifers during Mid-and/or Late-Gestation on Progeny Feedlot Performance and Carcass Characteristics. *Animals* **2022**, *12*, 588. [CrossRef]
14. McCoski, S.; Bradbery, A.; Marques, R.d.S.; Posbergh, C.; Sanford, C. Maternal Nutrition and Developmental Programming of Male Progeny. *Animals* **2021**, *11*, 2216. [CrossRef]
15. Copping, K.J.; Hoare, A.; Callaghan, M.; McMillen, I.C.; Rodgers, R.J.; Perry, V.E.A.; Copping, K.J.; Hoare, A.; Callaghan, M.; McMillen, I.C.; et al. Fetal Programming in 2-Year-Old Calving Heifers: Peri-Conception and First Trimester Protein Restriction Alters Fetal Growth in a Gender-Specific Manner. *Anim. Prod. Sci.* **2014**, *54*, 1333–1337. [CrossRef]
16. Maresca, S.; Valiente, S.L.; Rodriguez, A.M.; Testa, L.M.; Long, N.M.; Quintans, G.I.; Pavan, E. The Influence of Protein Restriction during Mid- to Late Gestation on Beef Offspring Growth, Carcass Characteristic and Meat Quality. *Meat Sci.* **2019**, *153*, 103–108. [CrossRef] [PubMed]
17. Novais, F.J.; Pires, P.R.L.; Alexandre, P.A.; Dromms, R.A.; Iglesias, A.H.; Ferraz, J.B.S.; Styczynski, M.P.W.; Fukumasu, H. Identification of a Metabolomic Signature Associated with Feed Efficiency in Beef Cattle. *BMC Genom.* **2019**, *20*, 8. [CrossRef]

18. Antonelo; Gómez, J.F.M.; Cônsolo, N.R.B.; Beline, M.; Colnago, L.A.; Schilling, W.; Zhang, X.; Suman, S.P.; Gerrard, D.E.; Balieiro, J.C.C.; et al. Metabolites and Metabolic Pathways Correlated With Beef Tenderness. *Meat Muscle Biol.* **2020**, *4*, 19–20. [CrossRef]
19. Keusch, G.T. What Do *-omics* Mean for the Science and Policy of the Nutritional Sciences? *Am. J. Clin. Nutr.* **2006**, *83*, 520S–522S. [CrossRef]
20. Vlaanderen, J.; Moore, L.E.; Smith, M.T.; Lan, Q.; Zhang, L.; Skibola, C.F.; Rothman, N.; Vermeulen, R. Application of OMICS Technologies in Occupational and Environmental Health Research; Current Status and Projections. *Occup. Environ. Med.* **2010**, *67*, 136–143. [CrossRef]
21. Kitano, H. Systems Biology: A Brief Overview. *Science* **2002**, *295*, 1662–1664. [CrossRef] [PubMed]
22. Shi-Kai, Y.; Run-Hui, L.; Hui-Zi, J.; Xin-Ru, L.; Ji, Y.; Lei, S.; Zhang, W.-D. Omics in Pharmaceutical Research: Overview, Applications, Challenges, and Future Perspectives. *Chin. J. Nat. Med.* **2015**, *13*, 3–21. [CrossRef]
23. Wishart, D.S.; Tzur, D.; Knox, C.; Eisner, R.; Guo, A.C.; Young, N.; Cheng, D.; Jewell, K.; Arndt, D.; Sawhney, S.; et al. HMDB: The Human Metabolome Database. *Nucleic Acids Res.* **2007**, *35*, D521–D526. [CrossRef]
24. Ashrafian, H.; Sounderajah, V.; Glen, R.; Ebbels, T.; Blaise, B.J.; Kalra, D.; Kultima, K.; Spjuth, O.; Tenori, L.; Salek, R.M.; et al. Metabolomics: The Stethoscope for the Twenty-First Century. *Med. Princ. Pract.* **2021**, *30*, 301–310. [CrossRef]
25. Liebal, U.W.; Phan, A.N.T.; Sudhakar, M.; Raman, K.; Blank, L.M. Machine Learning Applications for Mass Spectrometry-Based Metabolomics. *Metabolites* **2020**, *10*, 243. [CrossRef]
26. Pang, Z.; Chong, J.; Zhou, G.; De Lima Morais, D.A.; Chang, L.; Barrette, M.; Gauthier, C.; Jacques, P.É.; Li, S.; Xia, J. MetaboAnalyst 5.0: Narrowing the Gap between Raw Spectra and Functional Insights. *Nucleic Acids Res.* **2021**, *49*, W388–W396. [CrossRef] [PubMed]
27. Wu, G. *Amino Acids: Biochemistry and Nutrition*; CRC Press: Boca Raton, FL, USA, 2013; ISBN 9781439861905.
28. Hipkiss, A.R.; Gaunitz, F. Inhibition of Tumour Cell Growth by Carnosine: Some Possible Mechanisms. *Amino Acids* **2014**, *46*, 327–337. [CrossRef] [PubMed]
29. Boldyrev, A.A.; Aldini, G.; Derave, W. Physiology and Pathophysiology of Carnosine. *Physiol. Rev.* **2013**, *93*, 1803–1845. [CrossRef] [PubMed]
30. Barca, A.; Ippati, S.; Urso, E.; Vetrugno, C.; Storelli, C.; Maffia, M.; Romano, A.; Verri, T. Carnosine Modulates the Sp1-Slc31a1/Ctr1 Copper-Sensing System and Influences Copper Homeostasis in Murine CNS-Derived Cells. *Am. J. Physiol. Cell Physiol.* **2019**, *316*, C235–C245. [CrossRef]
31. Oppermann, H.; Alvanos, A.; Seidel, C.; Meixensberger, J.; Gaunitz, F. Carnosine Influences Transcription via Epigenetic Regulation as Demonstrated by Enhanced Histone Acetylation of the Pyruvate Dehydrogenase Kinase 4 Promoter in Glioblastoma Cells. *Amino Acids* **2019**, *51*, 61–71. [CrossRef]
32. Cônsolo, N.R.B.; Buarque, V.L.M.; Silva, J.; Poleti, M.D.; Barbosa, L.C.G.S.; Higuera-Padilla, A.; Gómez, J.F.M.; Colnago, L.A.; Gerrard, D.E.; Saran Netto, A.; et al. Muscle and Liver Metabolomic Signatures Associated with Residual Feed Intake in Nellore Cattle. *Anim. Feed Sci. Technol.* **2021**, *271*, 114757. [CrossRef]
33. Cui, J.; Pottosin, I.; Lamade, E.; Tcherkez, G. What Is the Role of Putrescine Accumulated under Potassium Deficiency? *Plant Cell Environ.* **2020**, *43*, 1331–1347. [CrossRef]
34. Tkachenko, A.G.; Pshenichnov, M.R.; Salakhetdinova, Y.; Nesterova, L.Y. The Role of Putrescine and Potassium Transport in the Regulation of DNA Topology during Escherichia Coti Adaptation to Heat Stress. *Mikrobiologiya* **1998**, *67*, 601–606.
35. Robert Michaud, M.; Benoit, J.B.; Lopez-Martinez, G.; Elnitsky, M.A.; Lee, R.E.; Denlinger, D.L. Metabolomics Reveals Unique and Shared Metabolic Changes in Response to Heat Shock, Freezing and Desiccation in the Antarctic Midge, Belgica Antarctica. *J. Insect Physiol.* **2008**, *54*, 645–655. [CrossRef]
36. Liao, Y.; Hu, R.; Wang, Z.; Peng, Q.; Dong, X.; Zhang, X.; Zou, H.; Pu, Q.; Xue, B.; Wang, L. Metabolomics Profiling of Serum and Urine in Three Beef Cattle Breeds Revealed Different Levels of Tolerance to Heat Stress. *J. Agric. Food Chem.* **2018**, *66*, 6926–6935. [CrossRef] [PubMed]
37. Isherwood, C.M.; Van der Veen, D.R.; Johnston, J.D.; Skene, D.J. Twenty-Four-Hour Rhythmicity of Circulating Metabolites: Effect of Body Mass and Type 2 Diabetes. *FASEB J.* **2017**, *31*, 5557–5567. [CrossRef]
38. Schmidt, J.A.; Fensom, G.K.; Rinaldi, S.; Scalbert, A.; Appleby, P.N.; Achaintre, D.; Gicquiau, A.; Gunter, M.J.; Ferrari, P.; Kaaks, R.; et al. Pre-Diagnostic Metabolite Concentrations and Prostate Cancer Risk in 1077 Cases and 1077 Matched Controls in the European Prospective Investigation into Cancer and Nutrition. *BMC Med.* **2017**, *15*, 122. [CrossRef]
39. Ghaffari, M.H.; Sadri, H.; Schuh, K.; Dusel, G.; Frieten, D.; Koch, C.; Prehn, C.; Adamski, J.; Sauerwein, H. Biogenic Amines: Concentrations in Serum and Skeletal Muscle from Late Pregnancy until Early Lactation in Dairy Cows with High versus Normal Body Condition Score. *J. Dairy Sci.* **2019**, *102*, 6571–6586. [CrossRef]
40. Humer, E.; Khol-Parisini, A.; Metzler-Zebeli, B.U.; Gruber, L.; Zebeli, Q. Alterations of the Lipid Metabolome in Dairy Cows Experiencing Excessive Lipolysis Early Postpartum. *PLoS ONE* **2016**, *11*, e0158633. [CrossRef]
41. Bischoff, S.C.; Barbara, G.; Buurman, W.; Ockhuizen, T.; Schulzke, J.D.; Serino, M.; Tilg, H.; Watson, A.; Wells, J.M. Intestinal Permeability—A New Target for Disease Prevention and Therapy. *BMC Gastroenterol.* **2014**, *14*, 189. [CrossRef]
42. Fragkos, K.C.; Forbes, A. Citrulline as a Marker of Intestinal Function and Absorption in Clinical Settings: A Systematic Review and Meta-Analysis. *UEG J.* **2018**, *6*, 181–191. [CrossRef]

43. Gilbreath, K.R.; Bazer, F.W.; Satterfield, M.C.; Wu, G. *Amino Acid Nutrition and Reproductive Performance in Ruminants*; Springer: Berlin/Heidelberg, Germany, 2021; Volume 1285.
44. Kott, M.L.; Pancini, S.; Speckhart, S.L.; Kimble, L.N.; White, R.R.; Stewart, J.L.; Johnson, S.E.; Ealy, A.D. Effects of Mid-Gestational l-Citrulline Supplementation to Twin-Bearing Ewes on Umbilical Blood Flow, Placental Development, and Lamb Production Traits. *Transl. Anim. Sci.* **2021**, *5*, txab102. [CrossRef] [PubMed]
45. Curis, E.; Nicolis, I.; Moinard, C.; Osowska, S.; Zerrouk, N.; Bénazeth, S.; Cynober, L. Almost All about Citrulline in Mammals. *Amino Acids* **2005**, *29*, 177–205. [CrossRef]
46. Cynober, L.; Moinard, C.; De Bandt, J.P. The 2009 ESPEN Sir David Cuthbertson. Citrulline: A New Major Signaling Molecule or Just Another Player in the Pharmaconutrition Game? *Clin. Nutr.* **2010**, *29*, 545–551. [CrossRef] [PubMed]
47. Uyanga, V.A.; Amevor, F.K.; Liu, M.; Cui, Z.; Zhao, X.; Lin, H. Potential Implications of Citrulline and Quercetin on Gut Functioning of Monogastric Animals and Humans: A Comprehensive Review. *Nutrients* **2021**, *13*, 3782. [CrossRef] [PubMed]
48. Gonzalez, A.M.; Trexler, E.T. Effects of Citrulline Supplementation on Exercise Performance in Humans: A Review of the Current Literature. *J. Strength Cond. Res.* **2020**, *34*, 1480–1495. [CrossRef]
49. Richard, D.M.; Dawes, M.A.; Mathias, C.W.; Acheson, A.; Hill-Kapturczak, N.; Dougherty, D.M. L-Tryptophan: Basic Metabolic Functions, Behavioral Research and Therapeutic Indications. *Int. J. Tryptophan Res.* **2009**, *2*, 45–60. [CrossRef]
50. Dougherty, D.M.; Marsh-Richard, D.M.; Mathias, C.W.; Hood, A.J.; Addicott, M.A.; Moeller, F.G.; Morgan, C.J.; Badawy, A.A.B. Comparison of 50- and 100-g L-Tryptophan Depletion and Loading Formulations for Altering 5-HT Synthesis: Pharmacokinetics, Side Effects, and Mood States. *Psychopharmacology* **2008**, *198*, 431–445. [CrossRef]
51. Marsh, D.M.; Dougherty, D.M.; Moeller, F.G.; Swann, A.C.; Spiga, R. Laboratory-Measured Aggressive Behavior of Women: Acute Tryptophan Depletion and Augmentation. *Neuropsychopharmacology* **2002**, *26*, 660–671. [CrossRef]
52. Rambali, B. *The Contribution of Cocoa Additive to Cigarette Smoking Addiction*; Rijksinstituut voor Volksgezondheid en Milieu RIVM: Bilthoven, The Netherlands, 2002.
53. Sainio, E.L.; Pulkki, K.; Young, S.N. L-Tryptophan: Biochemical, Nutritional and Pharmacological Aspects. *Amin. Acids* **1996**, *10*, 21–47. [CrossRef]
54. Fiore, A.; Murray, P.J. Tryptophan and Indole Metabolism in Immune Regulation. *Curr. Opin. Immunol.* **2021**, *70*, 7–14. [CrossRef] [PubMed]
55. Mohammad-Zadeh, L.F.; Moses, L.; Gwaltney-Brant, S.M. Serotonin: A Review. *J. Vet. Pharmacol. Ther.* **2008**, *31*, 187–199. [CrossRef] [PubMed]
56. Laporta, J.; Moore, S.A.E.; Peters, M.W.; Peters, T.L.; Hernandez, L.L. Short Communication: Circulating Serotonin (5-HT) Concentrations on Day 1 of Lactation as a Potential Predictor of Transition-Related Disorders. *J. Dairy Sci.* **2013**, *96*, 5146–5150. [CrossRef]
57. Laporta, J.; Moore, S.A.E.; Weaver, S.R.; Cronick, C.M.; Olsen, M.; Prichard, A.P.; Schnell, B.P.; Crenshaw, T.D.; Peñagaricano, F.; Bruckmaier, R.M.; et al. Increasing Serotonin Concentrations Alter Calcium and Energy Metabolism in Dairy Cows. *J. Endocrinol.* **2015**, *226*, 43–55. [CrossRef] [PubMed]
58. Donovan, M.H.; Tecott, L.H. Serotonin and the Regulation of Mammalian Energy Balance. *Front. Neurosci.* **2013**, *7*, 36. [CrossRef]
59. Watanabe, H.; Saito, R.; Nakano, T.; Takahashi, H.; Takahashi, Y.; Sumiyoshi, K.; Sato, K.; Chen, X.; Okada, N.; Iwasaki, S.; et al. Effect of Peripheral 5-HT on Glucose and Lipid Metabolism in Wether Sheep. *PLoS ONE* **2014**, *9*, e88058. [CrossRef]
60. Koopmans, S.J.; Guzik, A.C.; Van Der Meulen, J.; Dekker, R.; Kogut, J.; Kerr, B.J.; Southern, L.L. Effects of Supplemental L-Tryptophan on Serotonin, Cortisol, Intestinal Integrity, and Behavior in Weanling Piglets. *J. Anim. Sci.* **2006**, *84*, 963–971. [CrossRef]
61. Hernández-Castellano, L.E.; Özçelik, R.; Hernandez, L.L.; Bruckmaier, R.M. Short Communication: Supplementation of Colostrum and Milk with 5-Hydroxy-L-Tryptophan Affects Immune Factors but Not Growth Performance in Newborn Calves. *J. Dairy Sci.* **2018**, *101*, 794–800. [CrossRef]
62. Ma, H.; Zhang, W.; Song, W.H.; Sun, P.; Jia, Z.H. Effects of Tryptophan Supplementation on Cashmere Fiber Characteristics, Serum Tryptophan, and Related Hormone Concentrations in Cashmere Goats. *Domest. Anim. Endocrinol.* **2012**, *43*, 239–250. [CrossRef]
63. Hernández-Castellano, L.E.; Hernandez, L.L.; Weaver, S.; Bruckmaier, R.M. Increased Serum Serotonin Improves Parturient Calcium Homeostasis in Dairy Cows. *J. Dairy Sci.* **2017**, *100*, 1580–1587. [CrossRef]
64. Valente, E.E.L.; Klotz, J.L.; Ahn, G.; Harmon, D.L. Pattern of Postruminal Administration of L-Tryptophan Affects Blood Serotonin in Cattle. *Domest. Anim. Endocrinol.* **2021**, *74*, 106574. [CrossRef]
65. Li, S.; Gao, D.; Jiang, Y. Function, Detection and Alteration of Acylcarnitine Metabolism in Hepatocellular Carcinoma. *Metabolites* **2019**, *9*, 36. [CrossRef] [PubMed]
66. Batchuluun, B.; Al Rijjal, D.; Prentice, K.J.; Eversley, J.A.; Burdett, E.; Mohan, H.; Bhattacharjee, A.; Gunderson, E.P.; Liu, Y.; Wheeler, M.B. Elevated Medium-Chain Acylcarnitines Are Associated With Gestational Diabetes Mellitus and Early Progression to Type 2 Diabetes and Induce Pancreatic β-Cell Dysfunction. *Diabetes* **2018**, *67*, 885–897. [CrossRef] [PubMed]
67. Qu, Q.; Zeng, F.; Liu, X.; Wang, Q.J.; Deng, F. Fatty Acid Oxidation and Carnitine Palmitoyltransferase I: Emerging Therapeutic Targets in Cancer. *Cell Death Dis.* **2016**, *7*, e2226. [CrossRef]

68. Xiang, L.; Wei, J.; Tian, X.Y.; Wang, B.; Chan, W.; Li, S.; Tang, Z.; Zhang, H.; Cheang, W.S.; Zhao, Q.; et al. Comprehensive Analysis of Acylcarnitine Species in Db/Db Mouse Using a Novel Method of High-Resolution Parallel Reaction Monitoring Reveals Widespread Metabolic Dysfunction Induced by Diabetes. *Anal. Chem.* **2017**, *89*, 10368–10375. [CrossRef]
69. Seiler, S.E.; Koves, T.R.; Gooding, J.R.; Wong, K.E.; Stevens, R.D.; Ilkayeva, O.R.; Wittmann, A.H.; DeBalsi, K.L.; Davies, M.N.; Lindeboom, L.; et al. Carnitine Acetyltransferase Mitigates Metabolic Inertia and Muscle Fatigue during Exercise. *Cell Metab.* **2015**, *22*, 65–76. [CrossRef] [PubMed]
70. Casals, N.; Zammit, V.; Herrero, L.; Fadó, R.; Rodríguez-Rodríguez, R.; Serra, D. Carnitine Palmitoyltransferase 1C: From Cognition to Cancer. *Prog. Lipid Res.* **2016**, *61*, 134–148. [CrossRef] [PubMed]
71. Kim, H.I.; Raffler, J.; Lu, W.; Lee, J.J.; Abbey, D.; Saleheen, D.; Rabinowitz, J.D.; Bennett, M.J.; Hand, N.J.; Brown, C.; et al. Fine Mapping and Functional Analysis Reveal a Role of SLC22A1 in Acylcarnitine Transport. *Am. J. Hum. Genet.* **2017**, *101*, 489–502. [CrossRef]
72. Ladeira, M.M.; Schoonmaker, J.P.; Swanson, K.C.; Duckett, S.K.; Gionbelli, M.P.; Rodrigues, L.M.; Teixeira, P.D. Review: Nutrigenomics of Marbling and Fatty Acid Profile in Ruminant Meat. *Animal* **2018**, *12*, s282–s294. [CrossRef]
73. Nguyen, D.V.; Nguyen, O.C.; Malau-Aduli, A.E.O. Main Regulatory Factors of Marbling Level in Beef Cattle. *Vet. Anim. Sci.* **2021**, *14*, 100219. [CrossRef]
74. Pastenkos, G.; Miller, J.L.; Pritchard, S.M.; Nicola, A.V. Role of Sphingomyelin in Alphaherpesvirus Entry. *J. Virol.* **2019**, *93*, e01547-18. [CrossRef] [PubMed]
75. McNeil, P.L.; Steinhardt, R.A. Loss, Restoration, and Maintenance of Plasma Membrane Integrity. *J. Cell Biol.* **1997**, *137*, 1–4. [CrossRef] [PubMed]
76. Brown, D.A.; London, E. Functions of Lipid Rafts in Biological Membranes. *Annu. Rev. Cell Dev. Biol.* **1998**, *14*, 111–136. [CrossRef] [PubMed]
77. Rietveld, A.; Simons, K. The Differential Miscibility of Lipids as the Basis for the Formation of Functional Membrane Rafts. *Biochim. Biophys. Acta Rev. Biomembr.* **1998**, *1376*, 467–479. [CrossRef]
78. Hanamatsu, H.; Ohnishi, S.; Sakai, S.; Yuyama, K.; Mitsutake, S.; Takeda, H.; Hashino, S.; Igarashi, Y. Altered Levels of Serum Sphingomyelin and Ceramide Containing Distinct Acyl Chains in Young Obese Adults. *Nutr. Diabetes* **2014**, *4*, e141. [CrossRef] [PubMed]
79. McAllister, T.A.; Stanford, K.; Chaves, A.V.; Evans, P.R.; de Souza Figueiredo, E.E.; Ribeiro, G. Nutrition, Feeding and Management of Beef Cattle in Intensive and Extensive Production Systems. In *Animal Agriculture: Sustainability, Challenges and Innovations*; Academic Press: Cambridge, MA, USA, 2020; pp. 75–98, ISBN 9780128170526.
80. Summer, A.; Lora, I.; Formaggioni, P.; Gottardo, F. Impact of Heat Stress on Milk and Meat Production. *Anim. Front.* **2019**, *9*, 39–46. [CrossRef]

metabolites

Article

Effect of Cow-Calf Supplementation on Gene Expression, Processes, and Pathways Related to Adipogenesis and Lipogenesis in *Longissimus thoracis* Muscle of F1 Angus × Nellore Cattle at Weaning

Germán Darío Ramírez-Zamudio [1], Maria Júlia Generoso Ganga [2], Guilherme Luis Pereira [2,3], Ricardo Perecin Nociti [1], Marcos Roberto Chiaratti [4], Reinaldo Fernandes Cooke [5], Luis Artur Loyola Chardulo [2,3], Welder Angelo Baldassini [2,3], Otávio Rodrigues Machado-Neto [2,3] and Rogério Abdallah Curi [2,3,*]

1. College of Animal Science and Food Engineering, São Paulo University (USP), Pirassununga 13635-900, SP, Brazil
2. School of Agriculture and Veterinary Sciences (FCAV), São Paulo State University (UNESP), Jaboticabal 14884-900, SP, Brazil
3. School of Veterinary Medicine and Animal Science (FMVZ), São Paulo State University (UNESP), Botucatu 18618-681, SP, Brazil
4. Department of Genetics and Evolution, Federal University of São Carlos (UFSCAR), São Carlos 13565-905, SP, Brazil
5. Department of Animal Science, Texas A&M University, College Station, TX 77843, USA

* Correspondence: rogerio.curi@unesp.br

Abstract: The aim of this study was to identify differentially expressed genes, biological processes, and metabolic pathways related to adipogenesis and lipogenesis in calves receiving different diets during the cow-calf phase. Forty-eight uncastrated F1 Angus × Nellore males were randomly assigned to two treatments from thirty days of age to weaning: no creep feeding (G1) or creep feeding (G2). The creep feed offered contained ground corn (44.8%), soybean meal (40.4%), and mineral core (14.8%), with 22% crude protein and 65% total digestible nutrients in dry matter. After weaning, the animals were feedlot finished for 180 days and fed a single diet containing 12.6% forage and 87.4% corn-based concentrate. *Longissimus thoracis* muscle samples were collected by biopsy at weaning for transcriptome analysis and at slaughter for the measurement of intramuscular fat content (IMF) and marbling score (MS). Animals of G2 had 17.2% and 14.0% higher IMF and MS, respectively ($p < 0.05$). We identified 947 differentially expressed genes ($\log_2$ fold change 0.5, FDR 5%); of these, 504 were upregulated and 443 were downregulated in G2. Part of the genes upregulated in G2 were related to PPAR signaling (*PPARA*, *SLC27A1*, *FABP3*, and *DBI*), unsaturated fatty acid synthesis (*FADS1*, *FADS2*, *SCD*, and *SCD5*), and fatty acid metabolism (*FASN*, *FADS1*, *FADS2*, *SCD*, and *SCD5*). Regarding biological processes, the genes upregulated in G2 were related to cholesterol biosynthesis (*EBP*, *CYP51A1*, *DHCR24*, and *LSS*), unsaturated fatty acid biosynthesis (*FADS2*, *SCD*, *SCD5*, and *FADS1*), and insulin sensitivity (*INSIG1* and *LPIN2*). Cow-calf supplementation G2 positively affected energy metabolism and lipid biosynthesis, and thus favored the deposition of marbling fat during the postweaning period, which was shown here in an unprecedented way, by analyzing the transcriptome, genes, pathways, and enriched processes due to the use of creep feeding.

Keywords: *Bos indicus*; carcass; marbling; meat quality; nutrigenomics

1. Introduction

Different strategies have been used in beef cattle to increase intramuscular fat (IMF) deposition. Early weaning and supplementation during the cow-calf phase are nutritional management practices that can induce different metabolic adaptations when compared to conventionally weaned animals [1]. The approach known as creep feeding consists of supplementation during the cow-calf (lactation) phase with grains or forage to obtain heavier individuals at weaning, to reduce the time to carcass finishing for slaughter, and to allow the dam to rest [2,3].

It is well-known that animals can respond to different environmental/nutritional factors by exhibiting phenotypic plasticity as a result of changes in gene expression patterns [4]. Thus, animal characteristics can be modified by nutritional modulations, which means that dietary exposures can have consequences for growth and health [5]. New evidence is constantly emerging that alterations in gene expression and in the phenotype of individuals, as a result of nutritional stimuli, may be due to epigenetic factors, including DNA methylation [6–8]. These modifications in gene expression can occur at specific loci or on a genomic scale [9–13].

In *Bos taurus*, it was identified that glucose is the best precursor for the synthesis of fatty acids in the intramuscular adipose tissue and the only carbonic compound with a higher concentration of its carbons in the intramuscular adipose tissue compared to the subcutaneous tissue, and in addition, lower glucose incorporation in post-pubertal animals was also noted [14]. Thus, the adequate intake of concentrate during the lactation period, especially the one rich in starch (grains), favors the proliferation and growth of adipocytes. This is due to the increase in glucose metabolism, which can lead to increased marbling without, however, increasing fat in the viscera, which contributes to overall meat quality and higher carcass yield [15]. Furthermore, Lopez et al. [16] proposed that higher levels of crude protein in the diet increase the digestion and intestinal absorption of starch, and promote the elevation of plasma insulin and glucose, which, consequently, increases the deposition of IMF, by favoring precursors of fatty acids.

Brazil, the world's largest exporter of unprocessed beef for almost 20 years [17], has been witnessing the growing use of crossbred animals (*Bos taurus* × *Bos indicus*), especially Angus × Nellore. This strategy is used for the production of more tender meat with more IMF in an attempt to add value to the final product and to meet the demands of consumers who are willing to pay more for quality. IMF deposition in beef can positively influence sensory attributes such as flavor, juiciness, and tenderness [18]. In contrast, less IMF (marbling), which is observed in animals with a predominance of the *Bos indicus* genotype, compromises the sensory attributes of meat. This fact can be explained by the stimulation of fibrogenesis with declining intramuscular adipogenesis, with a consequent increase in connective tissue content [19]. However, the molecular mechanisms that control the growth of these tissues have not yet been completely elucidated.

Considering the lack of information about the changes in gene expression that occur in crossbred *Bos taurus* × *Bos indicus* calves supplemented during the cow-calf phase, this study aimed to evaluate the effect of creep feeding on gene expression, biological processes, and metabolic pathways related to adipogenesis and lipogenesis in *Longissimus thoracis* (LT) muscle biopsies collected at weaning from F1 Angus × Nellore cattle. Such data are of great economic importance for cattle production systems that aim to manipulate marbling and to obtain better quality meat to attend more demanding markets that pay for quality.

2. Materials and Methods

2.1. Animals

Forty-eight uncastrated (intact) F1 Angus × Nellore males born to the same Aberdeen Angus (*Bos taurus*) sire (half-siblings) were used.

The experiment (cow-calf phase) was conducted on a commercial farm located in the city of Anhembi, State of São Paulo, Brazil. The animals were submitted to two different treatments during most of the cow-calf phase (from 30 days of age to weaning—approximately 210 days), with 24 animals per treatment: group 1 (G1/T1)—no creep feeding (conventional weaning), and group 2 (G2/T2)—creep feeding. In the creep feeding system, the animals were kept with their mothers in the same paddock but had exclusive and free access to supplement corresponding to approximately 1% of body weight during all the cow-calf phase. The creep feed offered to animals of G2 (from 30 to 210 days of life) consisted of dry matter (22% crude protein and 65% total digestible nutrients) containing ground corn (44.8%), soybean meal (40.4%), and mineral core (14.8%).

After weaning (mean of 210 days), the animals of the 2 treatments were transferred to an experimental feedlot (Botucatu, São Paulo, Brazil), where they were housed in covered collective pens (three animals/pen with 10 m^2 per animal) for approximately 180 days. The two groups received the same diet, containing 12.6% forage and 87.4% corn-based concentrate. The diet was formulated with the RLM 3.3 software (Ração de Lucro Máximo–Maximum Profit Ration, Piracicaba, São Paulo, Brazil) [20] using the NRC Tropicalizado ESALQ system and consisted of corn, soybean meal, Tifton hay, sugarcane bagasse, urea, and vitamin–mineral supplement (Table S1). The diet was offered ad libitum, twice a day at 8 am and 4 pm.

The animals were weighed after a 16 h fast at the beginning of the cow-calf phase (initial weight—BWi), at the end of weaning (weaning weight—WW), and at the end of the feedlot period (final weight—BWf). The average daily weight gain 1 (ADG1: beginning of cow-calf phase to weaning) was calculated from BWi and WW, and ADG2 (weaning to the end of the feedlot period) from BWf and WW.

2.2. *Collection of Muscle Tissue at Weaning and Slaughter*

At weaning (210 days), fragments of the LT muscle were collected by biopsy from 12 animals randomly selected from each treatment (n = 24). For biopsy, the lumbar region was shaved, and a local anesthetic was subcutaneously administered. The biopsies were performed at the height of the 13th rib. After cleaning the biopsy site, a 1 cm incision was made with a scalpel and a sterile Bergstrom biopsy needle was used to obtain 1 g of muscle tissue. The sample was immediately transferred to liquid nitrogen and stored in an ultra-freezer at −80 °C.

After the feedlot period, the 48 animals were slaughtered using the cerebral concussion technique and sectioning of the jugular vein. The carcasses were identified, washed, and divided into two halves. The half-carcasses were weighed individually to obtain the hot carcass weight (HCW) and kept in a cold room for approximately 24 h at 1 °C.

After cooling, the carcasses were removed from the cold room and weighed. After weighing, the LT muscle of the left half-carcass was separated and backfat thickness (BFT) and rib eye area (REA) were evaluated between the 12th and 13th ribs before deboning. Beef samples (sirloin steaks) were collected between the 12th and 13th ribs (cranial direction) of the left half-carcass and used for the laboratory analysis of physicochemical quality attributes.

2.3. *Analysis of Meat Quality*

The following physicochemical meat quality attributes were analyzed in the 48 slaughtered animals (n = 24/treatment): marbling score (MS), total lipids/intramuscular fat percentage (IMF), and Warner–Blatzler shear force (WBSF). The MS was determined by a single, previously trained evaluator following the Brazil Beef Quality reference standards (https://www.brazilbeefquality.com/) (accessed on 12 June 2021). These standards are numbered from 1 to 11 and provide a point scale ranging from 100 to 1100 (adapted from AUS-MEAT [21]), where the closer to 1100, the more marbled the meat, and the closer to 100, the less marbled the meat. The IMF was determined by infrared spectroscopy using a FoodScan™ (Foss NIRSystems, Laurel, MD, USA). The procedure standardized by Shackelford et al. [22] was adopted to measure WBSF at 7 and 14 days of aging (WBSF7 and WBSF14).

2.4. *Statistical Analysis of Weight, Weight Gain, Carcass, and Meat Data*

The weight and weight gain data and carcass and meat traits (BWi, WW, ADG1, BWf, ADG2, HCW, REA, BFT, MS, IMF, WBSF7, and WBSF14) of the 48 samples (n = 24/group) were analyzed regarding the presence of outliers, homogeneity of variance, and normality of residuals. The data were expressed as means and their respective standard errors. Transformations of the variables to provide an approximation of the normal distribution were not necessary. The means of the groups were compared by the *t*-test to detect significant differences ($p < 0.05$). These analyses were performed using the R software v.4.2.1 (Vienna, Austria) [23].

2.5. Analysis of Differential Gene Expression

2.5.1. RNA Extraction and Sequencing

After the extraction of total RNA, 24 genomic libraries were prepared, which consisted of 12 samples from each group/treatment (G1 and G2) collected at weaning. Total RNA was extracted individually from 100 mg of LT muscle using TRIzol® (Life Technologies, Carlsbad, CA, USA), according to the manufacturer's instructions, and its quality was analyzed in a Bioanalyzer 2100® (Agilent, Santa Clara, CA, USA). A minimum RNA integrity number (RIN) ≥ 7 was adopted to ensure adequate quality of the total RNA. RNAs with a poly-A tail, mainly the mRNA, were purified from total RNA using oligo-dT beads. It is well-known that many eukaryotic non-coding RNAs can have a poly-A tail in intermediary or in mature forms [24] and, in this way, eventually be captured by oligo-dT bead columns.

The cDNA libraries of each sample were prepared and multiplexed using the TruSeq RNA Sample Preparation kit (Illumina, San Diego, CA, USA) from 2 μg of total RNA, according to the TruSeq RNA Sample Preparation kit v2 guide (Illumina). The Bioanalyzer 2100® (Agilent) was used to estimate the average size of the libraries and quantitative PCR (RT-qPCR) using the KAPA Library Quantification kit (KAPA Biosystems, Wilmington, MA, USA) to quantify them. Clustering and sequencing were performed in one lane using the HiSeq2500 kit v4 2 × 100 bp (Illumina) to produce 100 bp paired-end (PE) reads. The HiSeq 2500® sequencer (Illumina) was used to reach a minimum coverage of 16 million reads per sample.

2.5.2. Mapping of Sequences to the Reference Genome and Identification of Differentially Expressed Genes

The sequence data generated by the HiSeq System Illumina platform were converted to FastQ format and separated by library (multiplexed data) using the Casava 1.8.2 software (Illumina). The FastQC v.0.11.9 software [25] was used to analyze the quality of raw reads. Adapter sequences and low-quality sequences were removed using Fastp v.0.23.1 [26]. After this step, the quality of the reads was reassessed by combined visualization of all FastQC outputs using the MultiQC v.1.13 program [27] to confirm improvements in quality. Next, the reads were mapped to the bovine reference genome (*Bos taurus*—ARS-UCD1.3), available at: http://www.ensembl.org/Bos_taurus/Info/Index/ (accessed on 19 August 2021), using the STAR v.2.7.20 program [28]. Mapping was performed independently for each sample. For each library, a file with the .bam extension was generated, which contained the alignment of the fragments to the reference genome. Mapped reads were counted using featurecounts v.2.0.3 [29] and only PE reads mapped to a single position of the genome (uniquely mapped PE reads) and to known chromosomes were used for differential gene expression analysis.

Differential gene expression was compared between the different groups/treatments at weaning (i.e., G1 vs. G2). First, graphical principal component analysis (PCA) of the read counts normalized to counts per million (CPM) was performed using the factoextra package [30] of the R software [23] to divide the samples based on gene expression patterns, examining the level of similarity/dissimilarity between groups. Differentially expressed genes (DEGs) were identified using the method implemented in edgeR v.3.40.0 [31] of the R software [23] for parameter estimation by the maximum likelihood method. Generalized linear models were used, assuming a negative binomial distribution of the count data. For this purpose, size factors were computed by the trimmed mean of M-values (TMM) for each pair of samples and overdispersion parameters were estimated for each gene by the Cox–Reid method [32]. The expression of each gene was calculated as the mean expression in all samples of each group and is reported as the mean of the logarithmic function of CPM. The fold change was calculated as the logarithmic function of the ratio between the CPM of G2 and G1 for each gene. The *p*-value associated with the difference in gene expression between groups was obtained by the likelihood ratio test. The Benjamini–Hochberg procedure [33] was used to control the false discovery rate (FDR). A $\log_2$ fold change of 0.5 and significance adjusted to FDR < 0.05 were adopted to identify DEGs. The expression profiles of 13 constitutive genes were analyzed to evaluate the quality of sequencing. These genes were: *actin beta—ACTB; beta-2-microglobulin—B2M; glyceraldehyde-3-phosphate dehydrogenase—GAPDH; glucuronidase beta—GUSB; hydroxymethylbilane synthase—HMBS; hypoxanthine phosphoribosyltransferase 1—HPRT1; phosphoglycerate kinase 1—PGK1; peptidylprolyl isomerase A—PPIA; ribosomal protein L13a—RPL13A; ribosomal protein lateral stalk subunit P0—RPLP0; succinate dehydrogenase complex flavoprotein subunit A—SDHA; TATA-box binding protein—TBP; transferrin receptor—TFRC.*

2.5.3. Functional Analysis of Differentially Expressed Genes

To understand the functional role of the genes identified as DEGs between groups at weaning, lists of up- and down-regulated genes were used in over-representation analysis (ORA) of gene ontology terms (GO terms: biological processes) and metabolic pathways (Kyoto Encyclopedia of Genes and Genomes (KEGG) database) using the online Database for Annotation, Visualization, and Integrated Discovery (DAVID v.6.8; https://david.ncifcrf.gov/home.jsp) (accessed on 12 October 2021) [34]. Biological processes and metabolic pathways were defined as enriched in the presence of at least three genes in each pathway or process and a *p*-value < 0.05.

The functional annotation clustering function of DAVID and the ClueGO package [35] of Cytoscape v.3.9.1 were used to identify the relationships between enriched processes and pathways. Due to their relationships with each other and their potential relationships with intramuscular fat content in cattle, processes and pathways were highlighted. To identify/visualize the participation of DEGs in the different enriched and highlighted biological processes and pathways, a HeatMap was generated using the ComplexHeatMap package [36] of the R software. The ClueGO and CluePedia packages [37] of Cytoscape were used to generate networks of shared DEGs between KEGG pathways and biological processes (FDR < 5%).

The online STRING (Search Tool for the Retrieval of Interacting Genes/Proteins) database (version 11.5), which collects and integrates information on functional interactions between genes/proteins for a large number of organisms [38], was used to reveal and visualize functional interactions between the DEGs of biological processes and KEGG pathways identified as enriched by DAVID and highlighted due to their relationships visualized in ClueGO and potential relationships with intramuscular adipogenesis and lipogenesis. Protein–protein interactions (PPIs) with a confidence score > 0.4 (a commonly used threshold) and an FDR < 0.05 were considered and are shown in the graph. The Markov cluster algorithm (MCL), an unsupervised clustering algorithm for graphs based on simulation of stochastic flow, was used for the clustering of genes (nodes) using the default inflation parameter 3 of STRING.

The *Bos taurus* database was used for the analyses that employed the DAVID, ClueGO/CluePedia, and STRING programs/packages.

3. Results

3.1. Pre- and Post-Weaning Performance, Carcass and Meat Quality

Table 1 shows the body weight of the animals at the different time points, weight gain, and carcass-related information. Significant differences ($p < 0.05$) between groups/treatments were observed for WW and ADG1. Regarding WW, animals of G1 (no supplementation) were lighter and those of G2 (creep feeding) were heavier. The same situation was observed for ADG1, i.e., G2 > G1. Animals of G1 and G2 did not differ in terms of BWf, and likewise for ADG2 during the feedlot period. No significant differences in BWi or HCW were observed between groups.

Table 1. Means and respective standard errors of body weight at the beginning of the cow-calf phase, body weight at weaning, body weight at the end of the feedlot period, pre- and post-weaning average daily gain, and hot carcass yield obtained for the two treatments.

	BWi (kg)	WW (kg)	ADG1 (kg)	BWf (kg)	ADG2 (kg)	HCW (kg)
G1 [1]	61.29 ± 2.41	228.92 ± 5.07 [b]	0.93 ± 0.02 [b]	484.64 ± 5.96	1.36 ± 0.02	269.22 ± 8.23
G2 [1]	57.55 ± 2.61	243.57 ± 5.70 [a]	1.03 ± 0.03 [a]	491.85 ± 7.85	1.32 ± 0.03	273.62 ± 9.28

[1] G1: no creep feeding, G2: creep feeding, BWi: initial body weight, WW: weaning weight, ADG1: average daily gain between WW and BWi, BWf: final body weight, ADG2: average daily gain between BWf and WW, HCW: hot carcass weight. [a,b] Means followed by different superscript letters significantly differ ($p < 0.05$).

Table 2 shows the quality variables, which were measured or estimated objectively or subjectively in the carcasses or meat samples of animals by laboratory analysis. No significant differences ($p > 0.05$) between groups were observed for REA, WBSF7, or WBSF14. On the other hand, there were significant differences ($p < 0.05$) in BFT, IMF, and MS. Animals of G2 (creep feeding) exhibited a higher mean BFT than G1 animals (no supplementation). Regarding IMF, an objective measure for the assessment of marbling, G2 animals had a higher fat content/percentage than G1 animals. Visual differences in relation to the fat content between LT muscle samples collected from animals of G1 and G2 are shown in Figure S1. The MI, a subjective measure of marbling, was higher in G2 animals compared to G1.

Table 2. Means and respective standard errors of carcass backfat thickness and rib eye area, fat percentage, marbling index, and shear force at 7 and 14 days of aging obtained for the two treatments.

	BFT (mm)	IMF (%)	MS	REA (cm^2)	WBSF7 (kg)	WBSF14 (kg)
G1 [1]	10.61 ± 0.42 [b]	4.95 ± 0.20 [b]	321.50 ± 13.65 [b]	67.94 ± 1.16	4.52 ± 0.11	3.45 ± 0.11
G2 [1]	12.96 ± 0.83 [a]	5.80 ± 0.23 [a]	366.11 ± 12.39 [a]	65.50 ± 0.93	4.28 ± 0.12	3.42 ± 0.09

[1] G1: group 1, G2: group 2, BFT: backfat thickness, IMF: fat percentage, MS: marbling score, REA: rib eye area, WBSF7 and WBSF14: Warner–Bratzler shear force at 7 and 14 days post-mortem, respectively. [a,b] Means followed by different superscript letters significantly differ ($p < 0.05$).

3.2. *Analysis of Differential Gene Expression*

3.2.1. Concentration and Integrity of Total RNA

The mean total RNA concentration extracted from the 24 samples was 260.27 ng/µL. The 260/280 nm (nucleic acid/protein) and 260/230 nm (nucleic acid/extraction contaminants) ratios were approximately 1.9, a value considered to be adequate. The mean contamination with genomic DNA was 1.04% (range: 0.73% to 1.11%). The mean RIN was 7.6 (range: 7.0 to 8.0). Thus, the samples were intact (all RINs > 7) and free of contaminants.

3.2.2. RNA Sequencing and Mapping of Reads to the Reference Genome

A total of 241.2 million PE reads (2 × 100 bp) were obtained; of these, 230.5 million were uniquely mapped PE reads. The coverage achieved by sequencing was 37X (coverage for all transcripts of all samples). An average of 9.6 million uniquely mapped PE reads were obtained per sample, corresponding to 95.56% of all PE reads generated.

The reads were mapped to 27,607 genes (protein coding and non-coding). However, considering a count of uniquely mapped PE reads ≥ 3 in at least 12 samples, the total number of expressed genes, and thus used in the differential gene expression analyzed, was 16,604. The number of genes detected by functional category after application of a filter/threshold that excludes genes with a low count is presented in Figure S2. Following protein-coding genes (15,672), the largest number of sequenced genes were those encoding transcription factors (656) and non-coding genes such as long non-coding RNA (lncRNA), short nuclear RNA (snRNA), and micro-RNA (miRNA). The latter are related to the control of gene expression and processing of messenger RNA.

The number and percentage of transcripts/fragments aligned to the bovine reference genome identified in samples collected from non-creep-fed (G1) and creep-fed (G2) animals, which showed significant differences between means for marbling measurements—IMF and MS (G1 < G2, $p < 0.05$), are shown in Tables 3 and 4, respectively.

Table 3. Total number of generated PE reads aligned to the reference genome, and total number and percentage of uniquely mapped PE reads in samples collected from G1 animals (no creep feeding) at weaning.

Animal/Sample	No. of Generated PE Reads	No. of Mapped PE Reads	No. of Uniquely Mapped PE Reads	% of Uniquely Mapped PE Reads
1/RC1	11,679,440	11,506,606	11,203,820	95.93
2/RC2	10,814,468	10,674,013	10,406,095	96.22
3/RC3	9,853,062	9,675,863	9,402,096	95.42
4/RC4	10,773,912	10,497,727	10,223,070	94.89
5/RC5	10,407,984	10,276,901	10,003,083	96.11
6/RC6	9,827,847	9,684,304	9,411,731	95.77
7/RC7	10,068,925	9,867,642	9,628,738	95.63
8/RC8	9,866,300	9,636,103	9,384,361	95.12
9/RC9	9,846,398	9,631,769	9,327,987	94.74
10/RC10	9,854,664	9,722,289	9,476,470	96.16
11/RC11	9,880,030	9,722,399	9,479,344	95.94
12/RC12	12,482,586	12,229,115	11,918,101	95.48
Mean	10,446,301	10,260,394	9,988,741	95.62

Table 4. Total number of generated PE reads aligned to the reference genome, and total number and percentage of uniquely mapped PE reads in samples collected from G2 animals (creep feeding) at weaning.

Animal/Sample	No. of Generated PE Reads	No. of Mapped PE Reads	No. of Uniquely Mapped PE Reads	% of Uniquely Mapped PE Reads
13/RC13	9,479,645	9,250,146	9,033,774	95.30
14/RC14	10,640,508	10,341,889	10,068,523	94.62
15/RC15	9,553,313	9,373,266	9,141,017	95.68
16/RC16	9,333,592	9,207,412	8,986,543	96.28
17/RC17	9,129,520	9,022,713	8,783,636	96.21
18/RC18	10,213,502	10,046,519	9,797,829	95.93
19/RC19	10,152,415	10,000,483	9,757,874	96.11
20/RC20	10,134,139	9,981,276	9,736,555	96.08
21/RC21	9,651,481	9,260,150	9,018,951	93.45
22/RC22	8,606,966	8,434,895	8,233,201	95.66
23/RC23	9,538,157	9,359,896	9,132,297	95.74
24/RC24	9,435,252	9,215,591	8,995,317	95.34
Mean	9,655,708	9,457,853	9,223,793	95.53

The boxplots of the read counts normalized by the size factor showed that the distribution of quartiles was consistent between the samples of the two groups, indicating good quality of the sequencing data (Figure S3).

In relation to the expression profile of the constitutive genes, the expression was similar between the experimental groups (Figure 1).

PCA showed that the first two principal components explained more than 20% of the variation among samples (Figure 2). In addition, the formation of clearly distinct groups of samples was observed at weaning, indicating an evident difference in the expression of genes between treatments. This fact illustrates the effect of cow-calf supplementation (creep feeding) on gene expression at weaning.

3.2.3. Identification of Differentially Expressed Genes

Fragments of LT muscle obtained by biopsy were used to identify differences in global gene expression between G1 and G2 at weaning. A total of 947 DEGs were identified ($\log_2$ fold change < -0.5 or >0.5, FDR $< 5\%$) between groups at weaning; of these, 443 were downregulated and 504 were upregulated in G2 (creep feeding).

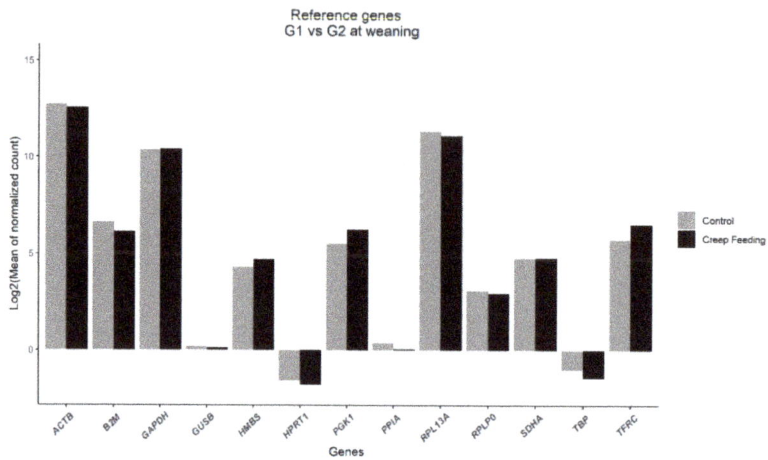

Figure 1. Expression profile of reference genes in the control group (G1, no creep feeding) and the group submitted to creep feeding (G2).

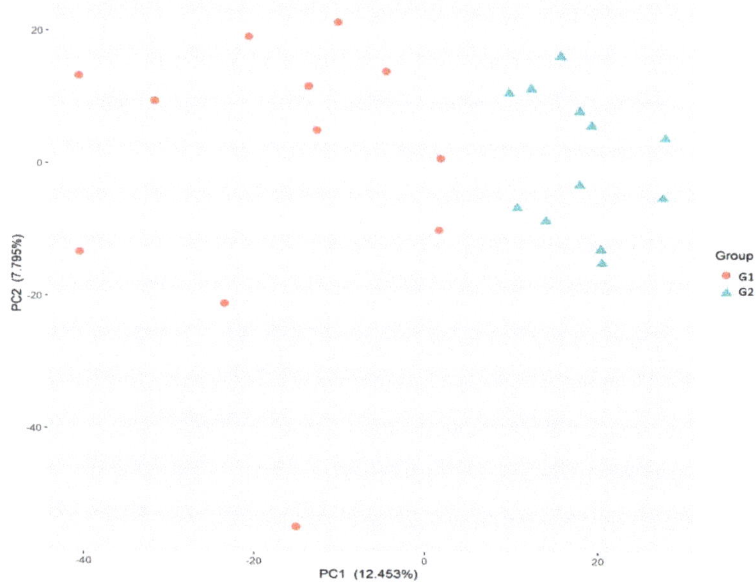

Figure 2. Principal component (PC) analysis performed based on normalized count data of gene expression for samples collected at weaning from G1 (control, no creep feeding) and G2 (creep feeding).

Figure 3 shows a Volcano plot that illustrates genes differentially expressed in G1 × G2 at weaning. The top 30 DEGs for comparison, with an adjusted *p*-value (FDR) < 0.05, are shown in Table 5. The complete list of the differentially expressed genes, with $\log_2$ fold change, *p*-value, and adjusted *p*-value of down- and up-regulated genes for comparison between groups at weaning, is provided in Table S2.

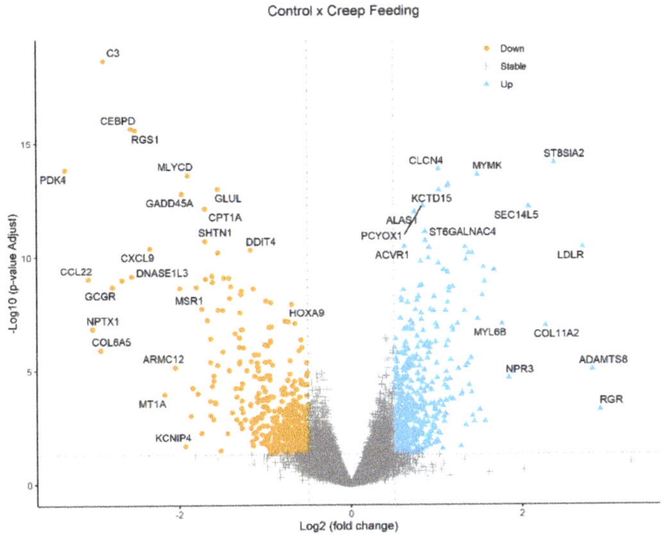

Figure 3. Volcano plot of $\log_2$ fold change (x-axis) versus -$\log_{10}$ p-value (FDR, y-axis) indicating differences in DEGs identified by the edgeR method (downregulated: $\log_2$ fold change < -0.5 and FDR < 0.05; upregulated: $\log_2$ fold change > 0.5 and FDR < 0.05) between G1 (control) × G2 (creep feeding) at weaning.

Table 5. Top 30 genes identified as differentially expressed between G1 (no creep feeding) × G2 (creep feeding) at weaning.

Gene ID Ensembl	Gene Symbol	Regulated [1]	$\log_2$ FC [2]	FDR [3]
ENSBTAG00000017280	*C3*	Down	−2.88	2.35×10^{-19}
ENSBTAG00000046307	*CEBPD*	Down	−2.56	2.31×10^{-16}
ENSBTAG00000021672	*RGS1*	Down	−2.51	2.70×10^{-16}
ENSBTAG00000048501	*ST8SIA2*	Up	2.37	8.32×10^{-15}
ENSBTAG00000011121	*CLCN4*	Up	1.02	1.52×10^{-14}
ENSBTAG00000014069	*PDK4*	Down	−3.34	1.52×10^{-14}
ENSBTAG00000004248	*MLYCD*	Down	−1.89	2.63×10^{-14}
ENSBTAG00000013242	*MYMK*	Up	1.47	2.67×10^{-14}
ENSBTAG00000002834	*CCDC69*	Up	1.13	6.93×10^{-14}
ENSBTAG00000022989	*FAM174B*	Up	1.12	8.62×10^{-14}
ENSBTAG00000013631	*GLUL*	Down	−1.54	1.08×10^{-13}
ENSBTAG00000017956	*KCTD15*	Up	1.02	1.27×10^{-13}
ENSBTAG00000013860	*GADD45A*	Down	−1.96	1.75×10^{-13}
ENSBTAG00000002783	*PCYOX1*	Up	0.83	6.15×10^{-13}
ENSBTAG00000007890	*SEC14L5*	Up	2.07	6.85×10^{-13}
ENSBTAG00000021999	*CPT1A*	Down	−1.70	7.95×10^{-13}
ENSBTAG00000004118	*ALAS1*	Up	0.74	1.13×10^{-12}
ENSBTAG00000046548	*ST6GALNAC4*	Up	0.86	8.42×10^{-12}
ENSBTAG00000007578	*SHTN1*	Down	−1.69	2.05×10^{-11}
ENSBTAG00000050158	———	Up	0.86	2.05×10^{-11}
ENSBTAG00000011909	*ACVR1*	Up	0.62	3.64×10^{-11}
ENSBTAG00000012314	*LDLR*	Up	2.71	4.10×10^{-11}
ENSBTAG00000015942	*DNAJA4*	Up	1.33	4.11×10^{-11}
ENSBTAG00000014265	*SREBF2*	Up	0.90	4.15×10^{-11}
ENSBTAG00000050852	*CXCL9*	Down	−2.34	4.23×10^{-11}
ENSBTAG00000000163	*DDIT4*	Down	−1.16	5.15×10^{-11}
ENSBTAG00000011437	———	Down	−1.54	6.69×10^{-11}
ENSBTAG00000016819	*FABP3*	Up	1.41	7.36×10^{-11}
ENSBTAG00000048728	———	Up	1.36	7.55×10^{-11}
ENSBTAG00000017280	*PMEPA1*	Up	0.90	1.19×10^{-10}

[1] Downregulated or upregulated in G2; [2] $\log_2$ fold change; [3] adjusted p-value (FDR) obtained by edgeR used to rank the differentially expressed genes presented in the table.

3.2.4. Functional Enrichment Analysis of Differentially Expressed Genes

Regarding the genes upregulated in G2 at weaning (n = 504), ORA identified 22 enriched metabolic KEGG pathways ($p < 0.05$) (Table S3). Among the pathways identified, five are highlighted due to their relationships with each other (identified by DAVID and ClueGO) and their potential relationships with adipose cell proliferation/adipogenesis and synthesis and degradation of intramuscular fat (lipolysis/lipogenesis) in cattle: PPAR signaling pathway (bta03320), steroid biosynthesis (bta00100), biosynthesis of unsaturated fatty acids (bta01040), apelin signaling pathway (bta04371), and fatty acid metabolism (bta01212). The upregulated DEGs in the enriched pathways and additional information are provided in Table 6. Regarding GO terms, 34 enriched biological processes were identified ($p < 0.05$) (Table S4). Four of these processes are highlighted due to their relationships with each other and potential relationships with intramuscular fat content in cattle: cholesterol biosynthetic process (GO:0006695), unsaturated fatty acid biosynthetic process (GO:0006636), sterol biosynthetic process (GO:0016126), and cellular response to insulin stimulus (GO:0032869). Table 7 shows the upregulated DEGs related to these processes.

Table 6. Enriched metabolic pathways ($p < 0.05$) highlighted in over-representation analysis of up- and down-regulated differentially expressed genes identified in G2 at weaning.

Term (KEGG)	Up/Down	No. of Genes	p-Value	Genes
PPAR signaling pathway	Up	9	<0.001	*FADS2, FABP3, SLC27A1, SCD, SCD5, AQP7, DBI, PPARA, RXRG*
Steroid biosynthesis	Up	5	<0.001	*SQLE, EBP, CYP51A1, DHCR24, LSS*
Biosynthesis of unsaturated fatty acids	Up	4	0.029	*FADS2, SCD, SCD5, FADS1*
Apelin signaling pathway	Up	8	0.039	*PRKAB2, SMAD3, PRKAA2, CCND1, MYL2, MYL3, APLNR, CALM3*
Fatty acid metabolism	Up	5	0.041	*FADS2, SCD, FASN, SCD5, FADS1*
AMPK signaling pathway	Down	8	0.023	*PFKFB4, CPT1A, PFKFB3, EIF4EBP1, MLYCD, CPT1B, FBP1, FOXO1*
Glucagon signaling pathway	Down	7	0.031	*CPT1A, GCGR, SIK1, CPT1B, FBP1, PLCB2, FOXO1*
PPAR signaling pathway	Down	6	0.044	*CPT1A, APOA1, ME3, ANGPTL4, CPT1B, PLIN5*

Table 7. Enriched biological processes ($p < 0.05$) highlighted in over-representation analysis of up- and down-regulated differentially expressed genes identified in G2 at weaning.

Term (GO_BP)	Up/Down	No. of Genes	p-Value	Genes
Cholesterol biosynthetic process	Up	5	0.002	*EBP, INSIG1, CYP51A1, DHCR24, LSS*
Unsaturated fatty acid biosynthetic process	Up	4	0.003	*FADS2, SCD, SCD5, FADS1*
Sterol biosynthetic process	Up	3	0.021	*SQLE, EBP, INSIG1*
Cellular response to insulin stimulus	Up	5	0.032	*GOT1, INSIG1, INHBB, LPIN2, HDAC9*
Positive regulation of lipid storage	Down	4	<0.001	*C3, IKBKE, FAM71F2, PLIN5*
Fatty acid metabolic process	Down	6	0.005	*C3, CPT1A, ACOT7, UCP3, HACL1, CPT1B*
Negative regulation of glycolytic process	Down	3	0.022	*DDIT4, NUPR1, FBP1*

Considering the genes downregulated in G2 at weaning (n = 443), 46 enriched pathways were identified ($p < 0.05$) (Table S5), and 3 of these pathways are highlighted due to the aforementioned relationships: AMPK signaling pathway (bta04152), glucagon signaling pathway (bta04922), and PPAR signaling pathway (bta03320). The downregulated DEGs participating in these pathways and additional information are provided in Table 6. Regarding GO terms, 52 enriched biological processes were identified ($p < 0.05$) (Table S6), and 3 are highlighted due to the aforementioned relationships: positive regulation of lipid storage (GO:0010884), fatty acid metabolic process (GO:0006631), and negative regulation of glycolytic process (GO:0045820). Table 7 shows the downregulated DEGs related to these processes.

Figure 4 shows the HeatMap that illustrates the relationship between each of the 52 up- and down-regulated DEGs identified in G2 at weaning and the 14 biological processes and metabolic pathways enriched in ORA, which are highlighted due to their relationships with each other and potential relationships with intramuscular fat content in cattle. This approach permitted to observe the participation of DEGs in one or more processes or pathways, as well as the magnitude of differences in gene expression between treatments.

Figure 4. Relationship between up- and down-regulated DEGs identified in G2 at weaning (x-axis) and enriched biological processes and pathways, which are highlighted due to their relationships with each other and possible relationships with intramuscular fat content in cattle (y-axis). The color scale indicating the $\log_2$ fold change is given on the right side of the figure.

Networks of shared up- and down-regulated DEGs in G2 at weaning between biological processes (GO_BP) and KEGG pathways, which were highlighted in ORA and due to the potential relationship with intramuscular fat/marbling in *Bos taurus*, are shown in Figures 5 and 6, respectively.

Figure 7 shows the network of PPI of 29 DEGs upregulated in G2 that participate in the biological processes (GO_BP) and metabolic pathways (KEGG) identified as enriched and highlighted in the G1 × G2 comparison at weaning due to their relationships with each other and potential relationships with intramuscular adipogenesis and lipogenesis. Sixty-nine edges (significant interactions) and six clusters consisting of eight to two genes were identified. The red and yellow clusters were the largest, with eight and five genes, respectively. Three clusters (dark green, light blue, and dark blue) contained only two genes. The mean clustering coefficient was 0.512. Six DEGs were disconnected from the network.

Figure 8 shows the network of PPI of 23 DEGs downregulated in G2 that participate in the biological processes (GO_BP) and metabolic pathways (KEGG) identified as enriched and highlighted in the G1 × G2 comparison at weaning due to their relationships with each other and potential relationships with intramuscular adipogenesis and lipogenesis. Twenty-three edges (significant interactions) and four clusters consisting of eight to two genes were identified. The red, yellow, and dark green clusters were the largest, with eight, three, and three genes, respectively. One of the clusters (dark blue) contained only two genes. The mean clustering coefficient was 0.467. Seven DEGs were disconnected from the network.

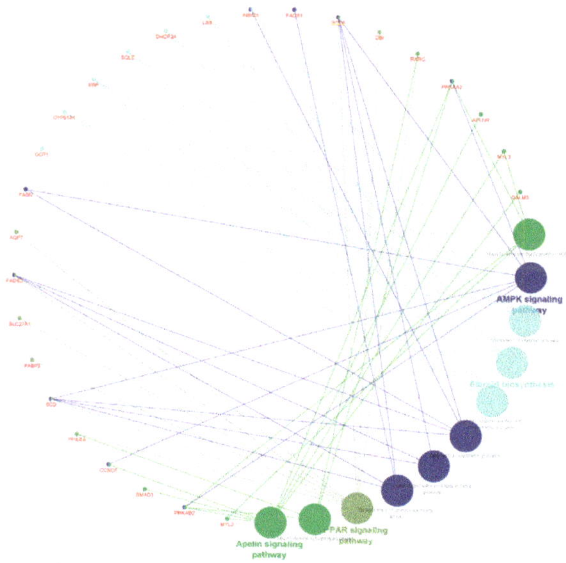

Figure 5. Circular network of shared upregulated DEGs in G2 at weaning between biological processes and KEGG pathways, which were highlighted in ORA and due to their potential relationship with intramuscular fat in cattle.

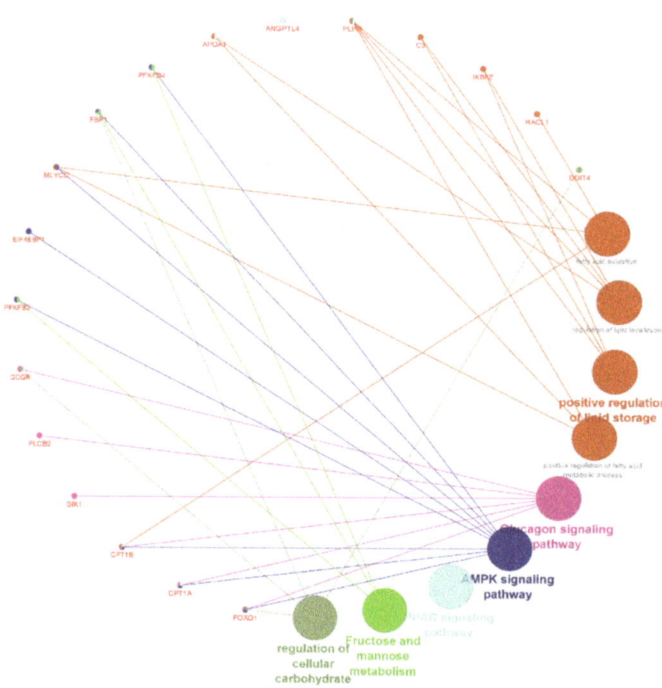

Figure 6. Circular network of shared downregulated DEGs in G2 at weaning between biological processes and KEGG pathways, which were highlighted in ORA and due to their potential relationship with intramuscular fat in cattle.

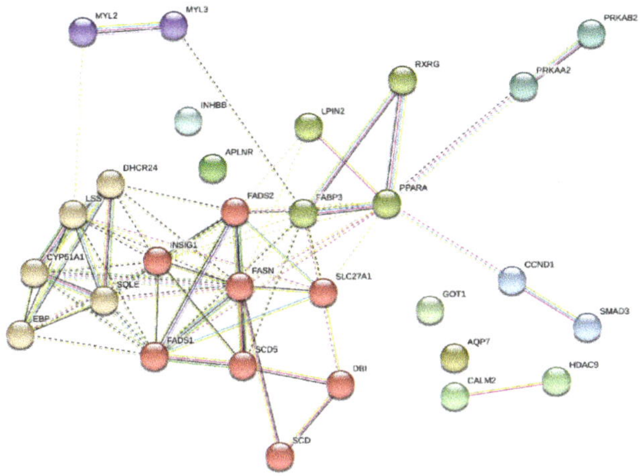

Figure 7. Protein–protein interaction network of DEGs upregulated in G2 that participate in the biological processes and KEGG pathways identified as enriched and highlighted in the G1 × G2 comparison at weaning. Type of interaction (edge color) between nodes (DEGs): light green—text mining; dark green—neighborhood; light blue—curated databases; dark blue—co-occurrence; pink—experiments; purple—protein homology; black—co-expression. A larger number of edges indicate stronger evidence/greater strength of the interaction between two nodes. The nodes that make up the clusters formed (joined by dotted edges) are illustrated with different colors.

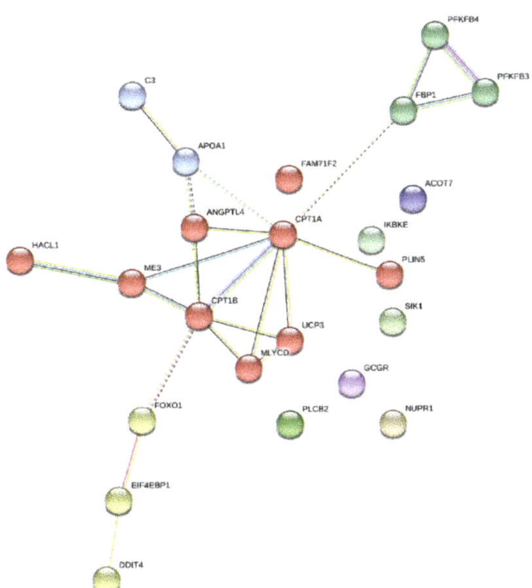

Figure 8. Protein–protein interaction network of DEGs downregulated in G2 that participate in the biological processes and KEGG pathways identified as enriched and highlighted in the G1 × G2 comparison at weaning. Type of interaction (edge color) between nodes (DEGs): light green—text mining; dark green—neighborhood; light blue—curated databases; dark blue—co-occurrence; pink—experiments; purple—protein homology; black—co-expression. A larger number of edges indicate stronger evidence/greater strength of the interaction between two nodes. The nodes that make up the clusters formed (joined by dotted edges) are illustrated with different colors.

4. Discussion

4.1. Pre- and Post-Weaning Performance, Carcass and Meat Quality

In tropical breeding systems, milk production of Nellore cows, which form the basis of the Brazilian cattle herd, no longer meets the calf's requirements for expression of its growth potential after the third month of lactation [39]. In F1 Angus × Nellore cattle, this interval is believed to be even shorter. Thus, obtaining the nutrients necessary to meet the growth requirements during lactation increasingly depends on the forage consumed by the animal. An important fact of the pasture-based systems in Brazil is that, in addition to the decline in the lactation curve of cows, the mass and nutritive value of pastures decrease due to seasonality, while the calf's nutritional requirements increase with the progression of growth [40]. Within this context, creep feeding is a strategy used to compensate the nutrient deficits of milk and forage, in addition to stimulating greater muscle development in the animal, which results in a shorter time to slaughter and improvement in carcass quality [41]. However, due to issues related to the offer of supplement, high supplies can affect fiber digestion from pasture, with negative consequences for feed efficiency [42,43] and calf performance [44]. This fact was not observed in the present study; on the contrary, calves that received 1% of their body weight as a supplement exhibited an additional daily weight gain of approximately 100 g and an additional total gain of 14.6 kg at weaning when compared to non-supplemented animals. However, in a meta-analysis conducted by Carvalho et al. [38], the additional daily weight gain was about 200 g and the total gain at weaning was 30 kg in male calves supplemented between 3 and 8 months of age when compared to non-supplemented animals. In the present study, supplementation during the lactation period did not exert any long-term effect on the performance of animals during the finishing period (BWf, ADG2, and HCW; Table 1). This finding suggests that the lack of difference in the efficiency of animals during the postweaning period may be related to gain composition [45], i.e., an early increase in the fat deposition rate may reduce the rate of lean tissue deposition [46]; consequently, weight gain in subsequent phases may be moderate. Similarly, calves supplemented during the lactation period showed a higher degree of carcass finishing (BFT) and greater fat deposition in beef, while no differences in weight gain were observed (Table 2).

4.2. Differentially Expressed Genes and Alterations in Biological Processes and Metabolic Pathways

Intramuscular fat is a desirable characteristic in some niche markets because of its positive effect on flavor, juiciness, and greater consumer perception of meat tenderness [47]. Although this quality trait is attributed to intense cell proliferation during the fetal period, there is a period of postnatal life within the first 250 days, known as the "marbling window", during which a high-grain diet can specifically lead to the recruitment of intramuscular adipocytes (adipocyte hyperplasia) that provide sites for fat deposition during finishing [48].

Studies indicate that the PPAR signaling pathway is important for the regulation of cell differentiation, energy balance, and lipid metabolism [49,50]. During lipogenesis, activation of this pathway upregulates the expression of *FABP*, *FASN*, and *SCD* [51,52]. The last two genes are important for de novo fatty acid synthesis [53] and were upregulated in creep-fed animals in the present study. Furthermore, Ward et al. [54] reported that, in addition to the upregulation of *FASN*, an increase in the level of marbling in beef is associated with the expression of Δ-6 and Δ-5 desaturases, enzymes that are regulated by *PPARA* [55]. This confirms that *PPARA* also acts on lipogenesis in skeletal muscle [56]. The Δ-5 and Δ-6 desaturases encoded by the *FADS1* and *FADS2* genes, respectively, belong to the fatty acid desaturase family, key proteins in the first desaturation reaction for endogenous formation of polyunsaturated fatty acids from dietary essential fatty acids (linoleic and linolenic) [57]. Linoleic and linolenic acids are unsaturated fatty acids found in oilseeds such as soybeans [58]. Although these fatty acids may undergo high ruminal biohydrogenation, part of them may escape into the duodenum and be absorbed [59]. In the present study, consumption of approximately 375 g of soybean meal per day is estimated in animals supplemented with 1% of body weight. Thus, part of the unsaturated fatty acids consumed from the supplement may have reached the skeletal muscle, with consequent upregulation of *FADS1* and *FADS2*.

Although calf supplementation during the lactation period had positive effects on the PPAR signaling pathway in lipid metabolism, downregulation of this pathway can downregulate the *APOA1* and *ANGPTL4* genes [60]. The *APOA1* gene encodes the structural and functional protein component of high-density lipoprotein (HDL), which promotes the reverse flow of cholesterol from tissues to excretion in the liver [61]. In the present study, downregulation of *APOA1* in supplemented calves may be a long-term effect of unsaturated fatty acid intake from soybean meal. Some studies have shown that the consumption of unsaturated fatty acids promotes a decrease in the expression of *APOA1* [62,63]. On the other hand, angiopoietin-like 4 encoded by the *ANGPTL4* gene mediates the inactivation of lipoprotein lipase involved in lipid metabolism [64], thereby inhibiting the uptake of triglycerides in adipocytes [65]. Downregulation of *APOA1* and *ANGPTL4* expression in supplemented calves suggests the accumulation of fatty acids, probably controlled by lipoprotein lipase that reduces the efflux of triglycerides from adipose tissue.

Calf supplementation at 1% of body weight during the lactation period positively influenced the lipogenic program by regulating genes related to the biosynthesis (*FASN*, *SCD*, *SCD5*, *FADS1*, and *FADS2*) and uptake of fatty acids (*FABP3*), while reducing the expression of transcription factors related to β-oxidation (*CPT1A*, *CPT1B*, and *UCP3*). A similar result was reported in a study on yaks, herbivores of the genus *Bos*, in which supplementation during the growth period stimulated de novo synthesis of fatty acids by upregulating *SREBF1*, *ACACA*, *FASN*, and *SCD1*, as well as the transcription factor *H-FABP* (*FABP3*), and downregulating *CPT1* [66]. Studies with rodents concluded that increased intracellular glucose availability inhibits CPT1 due to elevated malonyl-CoA concentrations which, in turn, reduce fatty acid oxidation [67,68]. In the present study, calves fed via creep feeding ingested high amounts of starch (>500 g/day), which can be transformed into glucose. Therefore, a high amount of glucose reaching the muscle tissue of these animals may have reduced the expression of genes related to fatty acid oxidation (*CPT1A*, *CPT1B*, and *UCP3*).

We also observed the regulation of genes that participate in the AMPK pathway of energy metabolism. However, the upregulation of the *PRKAA2* and *PRKAB2* genes, which encode the catalytic subunits necessary for AMPK activation during fatty acid oxidation [69], observed in the present study in supplemented calves was contradictory. A comparison between Angus and crossbred Angus × Simmental cattle, with the former exhibiting greater marbling potential, showed lower expression of these AMPK regulatory genes [70]. However, a study using cardiac muscle cells of mice indicated that AMPK acts as a downstream signaling molecule of apelin [71]. The latter plays an important role in energy metabolism by improving insulin sensitivity [72]. Another important enzyme in energy metabolism and fatty acid metabolism regulated by AMPK is malonyl-CoA decarboxylase [73]. This mitochondrial enzyme encoded by the *MLYCD* gene, which is responsible for increasing fatty acid oxidation by converting malonyl-CoA to acetyl-CoA [74], was downregulated in supplemented calves (Table 5). Studies on humans [74] and mice [75] have shown that silencing of this gene in muscle tissue increases the concentration of malonyl-CoA, thus increasing the utilization of glucose while reducing fatty acid oxidation. This fact may explain the downregulation of genes related to β-oxidation (*CPT1A*, *CPT1B*, and *UCP3*) in supplemented calves and may represent a protection mechanism against the development of diet-induced insulin resistance [76].

Higher expression of the complement 3 (*C3*) gene has been demonstrated in individuals with insulin resistance or hyperinsulinemia [77,78]. In the present study, calves supplemented at 1% of body weight showed downregulation of *C3* despite higher consumption of rapidly fermentable carbohydrates (starch). This finding may suggest that the insulin signaling cascade in muscle tissue is not dysregulated by high-carbohydrate intake. In addition to this reasoning, the activation of the apelin signaling pathway by upregulation of *PRKAA2* and *PRKAB2* and downregulation of *MLYCD* in supplemented calves suggests that these animals utilize more glucose as an energy source instead of fatty acids, a fact that may be closely related to changes in muscle fiber metabolism.

Skeletal muscle exhibits a certain metabolic plasticity, which allows to change substrate utilization (fat or glucose) for ATP production depending on the growth pattern of the animal [79] or nutritional stimuli [80]. During periods of accelerated muscle growth, energy expenditure for intramuscular fat and protein deposition is expected to increase [70]. Supplementation with high amounts of carbohydrates (starch) during the suckling period of calves probably stimulates glycolytic metabolism in skeletal muscle. Thus, a reduction in the fatty acid oxidation pathway for ATP production may increase the accumulation of lipids in adipose tissue. Skeletal muscle consists of a heterogeneous group of fibers that contain different myosin heavy-chain isoforms used to identify contractile and metabolic activity of muscle cells [81]. Although these isoforms were not identified in the present study, the downregulation of *CPT1A, CPT1B,* and *UCP3* in supplemented animals suggests a lower proportion of oxidative (type I) fibers in skeletal muscle since these transcripts are abundant in mitochondria [82]. According to Gagaoua and Picard [79], glycolytic (type II β) fibers contain a smaller number of mitochondria. These fibers are more abundant in muscles of animals with accelerated growth [83].

Fatty acid transport proteins (FATPs) are a family of six isoforms (FATP1–6). The isoform encoded by the *FATP1* gene (also known as *SLC27A1*) is highly expressed in muscle fibers, adipocytes, and hepatocytes due to the high absorption and accelerated metabolism of fatty acids in these cells [84]. In the present study, upregulation of *SLC27A1* was observed in skeletal muscle collected at weaning from animals supplemented during the cow-calf phase, and these animals thus exhibited higher IMF and MS (Table 2). These results agree with other studies on cattle that showed upregulation of *SLC27A1* in animals with higher intramuscular fat deposition [70,85]. Despite evidence that *FATP1* (*SLC27A1*) expression in muscle tissue is significantly associated with lipid accumulation [70,85–87], the results are contradictory since some studies have shown the opposite [88,89]. This effect on lipid metabolism (oxidation or esterification) can be explained by the localization of FATP1 that varies according to cell type and is tissue-specific [84]. For example, in muscle cells, FATP1 is most abundant in mitochondria since it is the key protein for supplying energy from fatty acid oxidation [89]. On the other hand, the cytoplasm is the site of highest abundance of FATP1 in adipocytes. When stimulated with insulin, these transporters are translocated to the plasma membrane for uptake, esterification, and accumulation of lipids [90]. An increase in FATP1 in adipose tissue increases the clearance of triglycerides from the vascular bed of muscle, thus reducing the availability of this substrate to be broken down into fatty acids [91]. Considering the negative effect of fatty acids on insulin-mediated glucose metabolism, a reduction in the flow of this substrate into muscle fibers may improve insulin sensitivity [91]. Since skeletal muscle is a heterogeneous tissue composed of muscle fibers, adipocytes, and fibroblasts, among other cell groups, the upregulated *SLC27A1* (*FATP1*) in the supplemented animals of the present study may be derived from adipocytes and not from muscle fibers. Furthermore, lower expression of *SLC27A1* in muscle fibers has been shown to be related to downregulation of *CPT1A*. Consequently, a lower rate of fatty acid oxidation in muscle fibers leads to a greater accumulation of intramuscular fat [88], if upregulation of *SLC27A1* (*FATP1*) in supplemented calves of the present study occurred in intramuscular adipose tissue. The supply of energy produced by the uptake and oxidation of fatty acids to muscle fibers may be reduced, a fact that partly explains the downregulation of pro-oxidative genes (*CPT1A, CPT1B,* and *UCP3*) in muscle of these animals.

In addition to stimulating lipogenesis, *FATP1* (*SLC27A1*) participates in the differentiation of pre-adipocytes regulated by *PPAR* [86,92]. During adipogenesis, *FATP1* increases the release of triglycerides into plasma and promotes the uptake of fatty acids by pre-adipocytes [93]. Once inside the cells, fatty acids bind to the ligand-binding domain that modifies the structure of PPAR, forming a heterodimer with the retinoid X receptor (RXR) [56]. Finally, the PPAR/RXR complex binds to the specific promoter regions of the target genes, inducing or inhibiting their expression [94]. In the present study, animals supplemented by creep feeding consumed approximately 375 g of soybean meal (Table 1). Diets containing soybean meal are rich in oleic, linoleic, and linolenic acids [58], which are potent regulators of *PPARA* [56]. It is therefore possible that, during the marbling window (250 days of age), the upregulation of *PPARA* and *RXRG* in supplemented animals is an indicator of greater recruitment of intramuscular adipocytes (adipogenesis).

The *FOXO1* gene encodes a transcription factor of the forkhead box class O family and plays an important role in the regulation of glucose metabolism through insulin signaling, in fatty acid metabolism (oxidation), and in the recruitment of pre-adipocytes in adipogenesis [95,96]. The role of *FOXO1* in adipogenesis may be either a promotor or an inhibitor depending on the stage of differentiation. Thus, in the post-mitotic stage, upregulation of *FOXO1* inhibits adipogenesis by activating the transcription of p21, a cell cycle inhibitor [97]. On the other hand, during the final differentiation and maturation of adipocytes (lipid metabolism), the inhibition of *FOXO1* impairs its binding to *PPAR*γ, which leads to the formation of the *PPAR*γ–*RXR*α–DNA complex in the transcription program [97]. In the present study, downregulation of *FOXO1* in calves supplemented during the lactation period suggests that the intramuscular adipose tissue collected at weaning may be in the final stage of differentiation and that adipocytes were ready to start fat accumulation during the finishing phase. Within this context, a larger number of differentiated intramuscular adipocytes in supplemented animals may also support the upregulation of genes related to the biosynthesis of fatty acids and cholesterol discussed in the previous paragraphs.

The *INSIG1* gene encodes an endoplasmic reticulum membrane protein that regulates glucose homeostasis and provides a negative feedback mechanism for cholesterol biosynthesis and lipogenesis [70,98]. Murine models [99] indicated that *INSIG1* expression is activated relatively late by the regulation of *PPARG* and *SREBF1* during adipogenesis and that its role in adipose tissue is to block the release of SREBP1 into the endoplasmic reticulum [100]. SREBP1 is a key protein in the control of *ACACA*, *FASN*, and *SCD* gene expression during de novo fatty acid synthesis [101]. However, the present findings do not support this mechanism since, despite the upregulation of *INSIG1* in supplemented calves, higher abundance of *FASN*, *SCD*, and *SCD5* was also observed, which would explain the higher IMF and MS (Table 2). Similarly, in a study on Angus and crossbred Angus × Simmental cattle, upregulation of *INSIG1* was observed in Angus animals with higher IMF and MS [70]. This fact suggests that overexpression of *INSIG1* in ruminants acts as a pro-lipogenic factor [70] and not as an anti-lipogenic gene, as demonstrated in mouse models [99].

Lipin is a protein encoded by a family of genes (*LPIN1*, *LPIN2*, and *LPIN3*) that play a key role in the lipogenesis and energy metabolism of adipose and muscle tissue [102,103]. Studies on transgenic mice with lipin deficiency or exclusive overexpression in muscle have shown that the upregulation of lipin reduces the utilization of fatty acids as energy, increasing the expression of lipogenic genes in adipocytes. Thus, the expression of lipin reduces muscle tissue energy expenditure and fat oxidation, slightly increasing the obesity of individuals [104,105]. In addition to the effects on fat deposition, altered lipin expression in muscle and adipose tissue affects insulin sensitivity. Lipin deficiency promotes insulin resistance, probably as a consequence of low leptin and adiponectin levels and impaired glucose absorption [106,107]. In contrast, lipin overexpression in transgenic mice was found to increase insulin sensitivity, although adipose tissue mass was doubled compared to normal mice [104]. This finding demonstrates that lipin improves the efficiency of fatty acid storage in adipocytes and predicts ectopic deposition of lipids in muscle tissue and consequent insulin impairment [105], in agreement with the results of the present study. Another study on cattle reported that starch-rich diets increased the expression of *LPIN2* and that this expression was positively correlated with insulin sensitivity and therefore associated with greater adipogenesis [70].

5. Conclusions

Supplementation of F1 Angus × Nellore calves during the cow-calf (lactation) phase prepares intramuscular adipose tissue for fat deposition during the postweaning period. The contribution of fatty acids such as linoleic and linolenic acids and high amounts of starch through ingredients such as soybean meal and ground corn included in supplements via creep feeding largely affected gene expression in the LT muscle at weaning, including adipogenic and lipogenic genes. Therefore, the upregulation of these genes that participate in critical metabolic pathways and biological processes related to PPAR signaling in adipogenesis and the insulin response, biosynthesis of fatty acids, and cholesterol in lipogenesis, contributes to the long-term intramuscular fat deposition during the finishing phase.

Supplementary Materials: The following supporting information can be downloaded at: https://www.mdpi.com/article/10.3390/metabo13020160/s1. Table S1: Ingredients and chemical-bromatological composition of the feedlot diet. Table S2: Complete list of differentially expressed genes (DEGs) between groups (G1 × G2) at weaning. Table S3: Enriched KEGG pathways for upregulated genes in group 2 at weaning. Table S4: Enriched biological processes (GO terms) for upregulated genes in group 2 at weaning. Table S5: Enriched KEGG pathways for downregulated genes in group 2 at weaning. Table S6: Enriched biological processes (GO terms) for downregulated genes in group 2 at weaning. Figure S1: Visual differences in relation to the fat content between *Longissimus thoracis* muscle samples collected from animals of G1 and G2. Figure S2: Number of genes by functional category detected by RNA sequencing considering the application of a filter that excludes genes with a low relative read count after normalization. Figure S3: Log2 boxplot of the read count normalized by the size factor per sample in G1 (control, no creep feeding) and G2 (creep feeding).

Author Contributions: Conceptualization, R.A.C., O.R.M.-N., L.A.L.C., G.L.P., W.A.B. and R.F.C.; methodology, O.R.M.-N., L.A.L.C., G.L.P., M.R.C. and R.A.C.; software, G.L.P. and R.P.N.; validation, G.L.P., W.A.B. and R.A.C.; formal analysis, G.L.P., L.A.L.C., W.A.B., M.R.C., R.P.N. and R.A.C.; investigation, G.D.R.-Z., M.J.G.G., G.L.P. and R.A.C.; resources, R.A.C. and O.R.M.-N.; data curation, R.A.C., O.R.M.-N. and L.A.L.C.; writing—original draft preparation, G.D.R.-Z., M.J.G.G. and R.A.C.; writing—review and editing, W.A.B., G.L.P., O.R.M.-N., L.A.L.C. and R.F.C.; visualization, W.A.B. and G.L.P.; supervision, R.A.C.; project administration, R.A.C. and O.R.M.-N.; funding acquisition, R.A.C. All authors have read and agreed to the published version of the manuscript.

Funding: This research was funded by the Research Foundation of the State of São Paulo (Fundação de Amparo à Pesquisa do Estado de São Paulo—FAPESP; grant number 2019/22621-3). This study was also financed in part by CAPES (Coordenação de Aperfeiçoamento de Pessoal de Nível Superior), Finance code 001.

Institutional Review Board Statement: The study was conducted in accordance with the Animal Use Ethics Committee of the School of Agriculture and Veterinary Sciences (FCAV), UNESP, Jaboticabal, São Paulo, Brazil (certificate number 013689/19).

Informed Consent Statement: Not applicable.

Data Availability Statement: All relevant data are presented within the paper. RNA-Seq data may be made available by contacting the corresponding author. Data is not publicly due to privacy.

Acknowledgments: The authors thank the São Paulo State University (Universidade Estadual Paulista, UNESP) for providing the infrastructure for all the stages of this research and the company Trouw Nutrition Brasil for providing part of the animals' diet.

Conflicts of Interest: The authors declare no conflict of interest.

References

1. Scheffler, J.; McCann, M.; Greiner, S.; Jiang, H.; Hanigan, M.; Bridges, G.; Lake, S.; Gerrard, D. Early metabolic imprinting events increase marbling scores in fed cattle. *J. Anim. Sci.* **2014**, *92*, 320–324. [CrossRef] [PubMed]
2. Dantas, C.C.O.; de Mattos Negrão, F.; Geron, L.J.V.; Mexia, A.A. O uso da técnica do Creep-feeding na suplementação de bezerros. *PUBVET* **2010**, *4*, 899–904.
3. Stewart, L. Creep feeding beef calves. *UGA Coop. Ext. Bull.* **2017**, *1315*, 1–8.

4. Zhang, N. Epigenetic modulation of DNA methylation by nutrition and its mechanisms in animals. *Anim. Nutr.* **2015**, *1*, 144–151. [CrossRef]
5. McKay, J.A.; Mathers, J.C. Diet induced epigenetic changes and their implications for health. *Acta Physiol.* **2011**, *202*, 103–118. [CrossRef]
6. Farias, N.; Ho, N.; Butler, S.; Delaney, L.; Morrison, J.; Shahrzad, S.; Coomber, B.L. The effects of folic acid on global DNA methylation and colonosphere formation in colon cancer cell lines. *J. Nutr. Biochem.* **2015**, *26*, 818–826. [CrossRef] [PubMed]
7. Farkas, S.A.; Befekadu, R.; Hahn-Strömberg, V.; Nilsson, T.K. DNA methylation and expression of the folate transporter genes in colorectal cancer. *Tumor Biol.* **2015**, *36*, 5581–5590. [CrossRef] [PubMed]
8. Day, J.J.; Kennedy, A.J.; Sweatt, J.D. DNA methylation and its implications and accessibility for neuropsychiatric therapeutics. *Annu. Rev. Pharmacol. Toxicol.* **2015**, *55*, 591. [CrossRef]
9. Vucetic, Z.; Kimmel, J.; Totoki, K.; Hollenbeck, E.; Reyes, T.M. Maternal high-fat diet alters methylation and gene expression of dopamine and opioid-related genes. *Endocrinology* **2010**, *151*, 4756–4764. [CrossRef]
10. Bogdarina, I.; Haase, A.; Langley-Evans, S.; Clark, A.J. Glucocorticoid effects on the programming of AT1b angiotensin receptor gene methylation and expression in the rat. *PLoS ONE* **2010**, *5*, e9237. [CrossRef]
11. Jousse, C.; Parry, L.; Lambert-Langlais, S.; Maurin, A.C.; Averous, J.; Bruhat, A.; Carraro, V.; Tost, J.; Letteron, P.; Chen, P. Perinatal undernutrition affects the methylation and expression of the leptin gene in adults: Implication for the understanding of metabolic syndrome. *FASEB J.* **2011**, *25*, 3271–3278. [CrossRef] [PubMed]
12. Dudley, K.J.; Sloboda, D.M.; Connor, K.L.; Beltrand, J.; Vickers, M.H. Offspring of mothers fed a high fat diet display hepatic cell cycle inhibition and associated changes in gene expression and DNA methylation. *PLoS ONE* **2011**, *6*, e21662. [CrossRef] [PubMed]
13. Altmann, S.; Murani, E.; Schwerin, M.; Metges, C.C.; Wimmers, K.; Ponsuksili, S. Somatic cytochrome c (CYCS) gene expression and promoter-specific DNA methylation in a porcine model of prenatal exposure to maternal dietary protein excess and restriction. *Br. J. Nutr.* **2012**, *107*, 791–799. [CrossRef]
14. Smith, S.B.; Crouse, J.D. Relative contributions of acetate, lactate and glucose to lipogenesis in bovine intramuscular and subcutaneous adipose tissue. *J. Nutr.* **1984**, *114*, 792–800. [CrossRef]
15. Baldassini, W.A.; Ferreira, M.S.; Santiago, B.M.; Chardulo, L.A.L.; Curi, R.A.; Lanna, D.P.D.; Ribeiro, R.V.; Martins, C.L.; Arrigoni, M.; Neto, O.R.M. Intake, performance, meat quality and fatty acid profile of Nellore bulls finished in feedlot with diets containing dry corn gluten feed. *Livest. Sci.* **2021**, *253*, 104715. [CrossRef]
16. Lopez, R.; Krehbiel, C.R.; Thomas, M.G.; Hallford, D.M.; Keisler, D.H.; Obeidat, B.S.; Hernandez, J.A.; Bryant, W.D.; Garcia, M.; Flores, R. Effect of increasing level of dietary protein on serum concentrations of metabolic hormones and mammary development in Holstein heifers consuming a moderate-energy diet. *J. Dairy Sci.* **2001**, *84*, 161.
17. ABIEC. Beef Report. 2022. Available online: https://www.abiec.com.br/publicacoes/beef-report-2022/ (accessed on 19 September 2022).
18. Hocquette, J.-F.; Gondret, F.; Baeza, E.; Medale, F.; Jurie, C.; Pethick, D. Intramuscular fat content in meat-producing animals: Development, genetic and nutritional control, and identification of putative markers. *Animal* **2010**, *4*, 303–319. [CrossRef]
19. Du, M.; Ford, S.P.; Zhu, M.-J. Optimizing livestock production efficiency through maternal nutritional management and fetal developmental programming. *Anim. Front.* **2017**, *7*, 5–11. [CrossRef]
20. Lanna, D.; Barioni, L.; Nepomuceno, N.; Almeida, R.; Tedeschi, L. *Ração de Lucro Máximo—RLM*, Version 3.2.1; RLM: Piracicaba, Brazil, 2011.
21. AUS-MEAT Limited. *Handbook of Australian Beef Processing*; AUS-MEAT Limited: Murarrie, Australia, 2018.
22. Shackelford, S.; Wheeler, T.; Koohmaraie, M. Evaluation of slice shear force as an objective method of assessing beef longissimus tenderness. *J. Anim. Sci.* **1999**, *77*, 2693–2699. [CrossRef]
23. R Core Team. *R: A Language and Environment for Statistical Computing*; R Foundation for Statistical Computing: Vienna, Austria; Available online: https://www.R-project.org/ (accessed on 2 June 2022).
24. Saini, H.K.; Griffiths-Jones, S.; Enright, A.J. Genomic analysis of human microRNA transcripts. *Proc. Natl. Acad. Sci. USA* **2007**, *104*, 17719–17724. [CrossRef]
25. Andrews, S. FastQC: A Quality Control Tool for High Throughput Sequence Data. Available online: http://www.bioinformatics.babraham.ac.uk/projects/fastqc/ (accessed on 2 June 2022).
26. Chen, S.; Zhou, Y.; Chen, Y.; Gu, J. Fastp: An ultra-fast all-in-one FASTQ preprocessor. *Bioinformatics* **2018**, *34*, i884–i890. [CrossRef] [PubMed]
27. Ewels, P.; Magnusson, M.; Lundin, S.; Käller, M. MultiQC: Summarize analysis results for multiple tools and samples in a single report. *Bioinformatics* **2016**, *32*, 3047–3048. [CrossRef] [PubMed]
28. Dobin, A.; Davis, C.A.; Schlesinger, F.; Drenkow, J.; Zaleski, C.; Jha, S.; Batut, P.; Chaisson, M.; Gingeras, T.R. STAR: Ultrafast universal RNA-seq aligner. *Bioinformatics* **2013**, *29*, 15–21. [CrossRef] [PubMed]
29. Liao, Y.; Smyth, G.K.; Shi, W. FeatureCounts: An efficient general purpose program for assigning sequence reads to genomic features. *Bioinformatics* **2014**, *30*, 923–930. [CrossRef]
30. Kassambara, A.; Mundt, F. Factoextra: Extract and Visualize the Results of Multivariate Data Analyses. Available online: https://CRAN.R-project.org/package=factoextra (accessed on 2 June 2022).

31. Robinson, M.D.; McCarthy, D.J.; Smyth, G.K. EdgeR: A Bioconductor package for differential expression analysis of digital gene expression data. *Bioinformatics* **2010**, *26*, 139–140. [CrossRef]
32. Robinson, M.D.; Oshlack, A. A scaling normalization method for differential expression analysis of RNA-seq data. *Genome Biol.* **2010**, *11*, R25. [CrossRef]
33. Benjamini, Y.; Hochberg, Y. Controlling the false discovery rate: A practical and powerful approach to multiple testing. *J. R. Stat. Soc. Ser. B Methodol.* **1995**, *57*, 289–300. [CrossRef]
34. Huang, D.W.; Sherman, B.T.; Tan, Q.; Kir, J.; Liu, D.; Bryant, D.; Guo, Y.; Stephens, R.; Baseler, M.W.; Lane, H.C. DAVID Bioinformatics Resources: Expanded annotation database and novel algorithms to better extract biology from large gene lists. *Nucleic Acids Res.* **2007**, *35*, W169–W175. [CrossRef]
35. Bindea, G.; Mlecnik, B.; Hackl, H.; Charoentong, P.; Tosolini, M.; Kirilovsky, A.; Fridman, W.-H.; Pagès, F.; Trajanoski, Z.; Galon, J. ClueGO: A Cytoscape plug-in to decipher functionally grouped gene ontology and pathway annotation networks. *Bioinformatics* **2009**, *25*, 1091–1093. [CrossRef]
36. Gu, Z.; Eils, R.; Schlesner, M. Complex heatmaps reveal patterns and correlations in multidimensional genomic data. *Bioinformatics* **2016**, *32*, 2847–2849. [CrossRef]
37. Bindea, G.; Galon, J.; Mlecnik, B. CluePedia Cytoscape plugin: Pathway insights using integrated experimental and in silico data. *Bioinformatics* **2013**, *29*, 661–663. [CrossRef] [PubMed]
38. Szklarczyk, D.; Morris, J.H.; Cook, H.; Kuhn, M.; Wyder, S.; Simonovic, M.; Santos, A.; Doncheva, N.T.; Roth, A.; Bork, P. The STRING database in 2017: Quality-controlled protein–protein association networks, made broadly accessible. *Nucleic Acids Res.* **2016**, *45*, gkw937. [CrossRef]
39. Costa e Silva, L.F.; Engle, T.E.; de Valadares Filho, S.; Rotta, P.P.; Villadiego, F.A.C.; Silva, F.A.S.; Martins, E.C.; Silva, L.H.R.; Paulino, M.F. Nellore cows and their calves during the lactation period: Performance, intake, milk composition, and total apparent digestibility. *Trop. Anim. Health Prod.* **2015**, *47*, 735–741. [CrossRef] [PubMed]
40. Fonseca, M.A.; Valadares Filho, S.D.C.; Henriques, L.T.; Paulino, P.V.R.; Detmann, E.; Fonseca, E.A.; Benedeti, P.D.B.; Silva, L.D.D. Nutritional requirements of nursing Nellore calves. *Rev. Bras. Zootec.* **2012**, *41*, 1212–1221. [CrossRef]
41. Carvalho, V.V.; Paulino, M.F.; Detmann, E.; Valadares Filho, S.C.; Lopes, S.A.; Rennó, L.N.; Sampaio, C.B.; Silva, A.G. A meta-analysis of the effects of creep-feeding supplementation on performance and nutritional characteristics by beef calves grazing on tropical pastures. *Livest. Sci.* **2019**, *227*, 175–182. [CrossRef]
42. Cremin, J., Jr.; Faulkner, D.; Merchen, N.; Fahey, G., Jr.; Fernando, R.; Willms, C. Digestion criteria in nursing beef calves supplemented with limited levels of protein and energy. *J. Anim. Sci.* **1991**, *69*, 1322–1331. [CrossRef]
43. Faulkner, D.; Hummel, D.; Buskirk, D.; Berger, L.; Parrett, D.; Cmarik, G. Performance and nutrient metabolism by nursing calves supplemented with limited or unlimited corn or soyhulls. *J. Anim. Sci.* **1994**, *72*, 470–477. [CrossRef]
44. Lopes, S.A.; Paulino, M.F.; Detmann, E.; Valente, É.E.L.; Rennó, L.N.; Valadares, R.F.D.; Cardenas, J.E.G.; de Almeida, D.M.; de Moura, F.H.; Oliveira, C.A.S. Evaluation of supplementation plans for suckling beef calves managed on tropical pasture. *Semin. Ciênc. Agrár.* **2017**, *38*, 1027–1039. [CrossRef]
45. Asher, A.; Shabtay, A.; Cohen-Zinder, M.; Aharoni, Y.; Miron, J.; Agmon, R.; Halachmi, I.; Orlov, A.; Haim, A.; Tedeschi, L. Consistency of feed efficiency ranking and mechanisms associated with inter-animal variation among growing calves. *J. Anim. Sci.* **2018**, *96*, 990–1009. [CrossRef]
46. Gerrits, W.J.; France, J.; Dijkstra, J.; Bosch, M.W.; Tolman, G.H.; Tamminga, S. Evaluation of a model integrating protein and energy metabolism in preruminant calves. *J. Nutr.* **1997**, *127*, 1243–1252. [CrossRef]
47. Hunt, M.; Garmyn, A.; O'Quinn, T.; Corbin, C.; Legako, J.; Rathmann, R.; Brooks, J.; Miller, M. Consumer assessment of beef palatability from four beef muscles from USDA Choice and Select graded carcasses. *Meat Sci.* **2014**, *98*, 1–8. [CrossRef]
48. Du, M.; Huang, Y.; Das, A.; Yang, Q.; Duarte, M.; Dodson, M.; Zhu, M.-J. Meat Science and Muscle Biology Symposium: Manipulating mesenchymal progenitor cell differentiation to optimize performance and carcass value of beef cattle. *J. Anim. Sci.* **2013**, *91*, 1419–1427. [CrossRef] [PubMed]
49. Gross, B.; Pawlak, M.; Lefebvre, P.; Staels, B. PPARs in obesity-induced T2DM, dyslipidaemia and NAFLD. *Nat. Rev. Endocrinol.* **2017**, *13*, 36–49. [CrossRef] [PubMed]
50. Chen, H.; Peng, T.; Shang, H.; Shang, X.; Zhao, X.; Qu, M.; Song, X. RNA-Seq Analysis Reveals the Potential Molecular Mechanisms of Puerarin on Intramuscular Fat Deposition in Heat-Stressed Beef Cattle. *Front. Nutr.* **2022**, *9*, 817557. [CrossRef]
51. Brown, J.D.; Plutzky, J. Peroxisome proliferator–activated receptors as transcriptional nodal points and therapeutic targets. *Circulation* **2007**, *115*, 518–533. [CrossRef] [PubMed]
52. Rangwala, S.M.; Lazar, M.A. Peroxisome proliferator-activated receptor γ in diabetes and metabolism. *Trends Pharmacol. Sci.* **2004**, *25*, 331–336. [CrossRef]
53. Ladeira, M.; Schoonmaker, J.; Gionbelli, M.; Dias, J.; Gionbelli, T.; Carvalho, J.R.; Teixeira, P. Nutrigenomics and beef quality: A review about lipogenesis. *Int. J. Mol. Sci.* **2016**, *17*, 918. [CrossRef]
54. Ward, R.; Woodward, B.; Otter, N.; Doran, O. Relationship between the expression of key lipogenic enzymes, fatty acid composition, and intramuscular fat content of Limousin and Aberdeen Angus cattle. *Livest. Sci.* **2010**, *127*, 22–29. [CrossRef]
55. Nakamura, M.T.; Nara, T.Y. Structure, function, and dietary regulation of (delta) 6,(delta) 5, and (delta) 9 desaturases. *Annu. Rev. Nutr.* **2004**, *24*, 345. [CrossRef]

56. Bionaz, M.; Chen, S.; Khan, M.J.; Loor, J.J. Functional Role of PPARs in Ruminants: Potential Targets for Fine-Tuning Metabolism during Growth and Lactation. *PPAR Res.* **2013**, *2013*, 28. [CrossRef]
57. Ibeagha-Awemu, E.M.; Akwanji, K.A.; Beaudoin, F.; Zhao, X. Associations between variants of FADS genes and omega-3 and omega-6 milk fatty acids of Canadian Holstein cows. *BMC Genet.* **2014**, *15*, 25. [CrossRef] [PubMed]
58. Ladeira, M.M.; de Oliveira, D.M.; Schoonmaker, J.P.; Chizzotti, M.L.; Barreto, H.G.; Paiva, L.V.; Coelho, T.C.; Neto, O.R.M.; Gionbelli, M.P.; Chalfun-Junior, A. Expression of lipogenic genes in the muscle of beef cattle fed oilseeds and vitamin E. *Agri Gene* **2020**, *15*, 100097. [CrossRef]
59. Duckett, S.; Gillis, M. Effects of oil source and fish oil addition on ruminal biohydrogenation of fatty acids and conjugated linoleic acid formation in beef steers fed finishing diets. *J. Anim. Sci.* **2010**, *88*, 2684–2691. [CrossRef] [PubMed]
60. Schroyen, M.; Li, B.; Arevalo Sureda, E.; Zhang, Y.; Leblois, J.; Deforce, D.; Van Nieuwerburgh, F.; Wavreille, J.; Everaert, N. Pre-Weaning Inulin Supplementation Alters the Ileal Transcriptome in Pigs Regarding Lipid Metabolism. *Vet. Sci.* **2021**, *8*, 207. [CrossRef] [PubMed]
61. Mangaraj, M.; Nanda, R.; Panda, S. Apolipoprotein AI: A molecule of diverse function. *Indian J. Clin. Biochem.* **2016**, *31*, 253–259. [CrossRef] [PubMed]
62. Poudyal, H.; Panchal, S.K.; Diwan, V.; Brown, L. Omega-3 fatty acids and metabolic syndrome: Effects and emerging mechanisms of action. *Prog. Lipid Res.* **2011**, *50*, 372–387. [CrossRef]
63. Fox, J.; McGill Jr, H.; Carey, K.; Getz, G. In vivo regulation of hepatic LDL receptor mRNA in the baboon. Differential effects of saturated and unsaturated fat. *J. Biol. Chem.* **1987**, *262*, 7014–7020. [CrossRef]
64. Blædel, T.; Holm, J.B.; Sundekilde, U.K.; Schmedes, M.S.; Hess, A.L.; Lorenzen, J.K.; Kristiansen, K.; Dalsgaard, T.K.; Astrup, A.; Larsen, L.H. A randomised, controlled, crossover study of the effect of diet on angiopoietin-like protein 4 (ANGPTL4) through modification of the gut microbiome. *J. Nutr. Sci.* **2016**, *5*, E45. [CrossRef]
65. Yoo, J.Y.; Kim, S.S. Probiotics and prebiotics: Present status and future perspectives on metabolic disorders. *Nutrients* **2016**, *8*, 173. [CrossRef]
66. Zhang, H.B.; Wang, Z.S.; Peng, Q.H.; Tan, C.; Zou, H.W. Effects of different levels of protein supplementary diet on gene expressions related to intramuscular deposition in early-weaned yaks. *Anim. Sci. J.* **2014**, *85*, 411–419. [CrossRef]
67. Turcotte, L.P.; Swenberger, J.R.; Yee, A.J. High carbohydrate availability increases LCFA uptake and decreases LCFA oxidation in perfused muscle. *Am. J. Physiol. Endocrinol. Metab.* **2002**, *282*, E177–E183. [CrossRef] [PubMed]
68. Yee, A.J.; Turcotte, L.P. Insulin fails to alter plasma LCFA metabolism in muscle perfused at similar glucose uptake. *Am. J. Physiol. Endocrinol. Metab.* **2002**, *283*, E73–E77. [CrossRef]
69. Zhang, M.; Zheng, D.; Peng, Z.; Zhu, Y.; Li, R.; Wu, Q.; Li, Y.; Li, H.; Xu, W.; Zhang, M. Identification of differentially expressed genes and lipid metabolism signaling pathways between muscle and fat tissues in broiler chickens. *J. Poult. Sci.* **2021**, *58*, 131–137. [CrossRef] [PubMed]
70. Graugnard, D.E.; Piantoni, P.; Bionaz, M.; Berger, L.L.; Faulkner, D.B.; Loor, J.J. Adipogenic and energy metabolism gene networks in longissimus lumborum during rapid post-weaning growth in Angus and Angus× Simmental cattle fed high-starch or low-starch diets. *BMC Genom.* **2009**, *10*, 142. [CrossRef] [PubMed]
71. Simpkin, J.C.; Yellon, D.M.; Davidson, S.M.; Lim, S.Y.; Wynne, A.M.; Smith, C.C. Apelin-13 and apelin-36 exhibit direct cardioprotective activity against ischemiareperfusion injury. *Basic Res. Cardiol.* **2007**, *102*, 518–528. [CrossRef]
72. Castan-Laurell, I.; Dray, C.; Knauf, C.; Kunduzova, O.; Valet, P. Apelin, a promising target for type 2 diabetes treatment? *Trends Endocrinol. Metab.* **2012**, *23*, 234–241. [CrossRef]
73. Ruderman, N.; Prentki, M. AMP kinase and malonyl-CoA: Targets for therapy of the metabolic syndrome. *Nat. Rev. Drug Discov.* **2004**, *3*, 340–351. [CrossRef]
74. Bouzakri, K.; Austin, R.; Rune, A.; Lassman, M.E.; Garcia-Roves, P.M.; Berger, J.P.; Krook, A.; Chibalin, A.V.; Zhang, B.B.; Zierath, J.R. Malonyl CoenzymeA decarboxylase regulates lipid and glucose metabolism in human skeletal muscle. *Diabetes* **2008**, *57*, 1508–1516. [CrossRef]
75. Dyck, J.R.; Hopkins, T.A.; Bonnet, S.; Michelakis, E.D.; Young, M.E.; Watanabe, M.; Kawase, Y.; Jishage, K.-I.; Lopaschuk, G.D. Absence of malonyl coenzyme A decarboxylase in mice increases cardiac glucose oxidation and protects the heart from ischemic injury. *Circulation* **2006**, *114*, 1721–1728. [CrossRef]
76. Koves, T.R.; Ussher, J.R.; Noland, R.C.; Slentz, D.; Mosedale, M.; Ilkayeva, O.; Bain, J.; Stevens, R.; Dyck, J.R.; Newgard, C.B. Mitochondrial overload and incomplete fatty acid oxidation contribute to skeletal muscle insulin resistance. *Cell Metab.* **2008**, *7*, 45–56. [CrossRef]
77. Weyer, C.; Tataranni, P.A.; Pratley, R.E. Insulin action and insulinemia are closely related to the fasting complement C3, but not acylation stimulating protein concentration. *Diabetes Care* **2000**, *23*, 779–785. [CrossRef]
78. Barbu, A.; Hamad, O.A.; Lind, L.; Ekdahl, K.N.; Nilsson, B. The role of complement factor C3 in lipid metabolism. *Mol. Immunol.* **2015**, *67*, 101–107. [CrossRef] [PubMed]
79. Picard, B.; Gagaoua, M. Muscle fiber properties in cattle and their relationships with meat qualities: An overview. *J. Agric. Food Chem.* **2020**, *68*, 6021–6039. [CrossRef] [PubMed]
80. Aragão, R.D.S.; Guzmán-Quevedo, O.; Pérez-García, G.; Manhaes-de-Castro, R.; Bolanos-Jiménez, F. Maternal protein restriction impairs the transcriptional metabolic flexibility of skeletal muscle in adult rat offspring. *Br. J. Nutr.* **2014**, *112*, 328–337. [CrossRef] [PubMed]

81. Schiaffino, S.; Reggiani, C. Fiber types in mammalian skeletal muscles. *Physiol. Rev.* **2011**, *91*, 1447–1531. [CrossRef] [PubMed]
82. Turner, N.; Hariharan, K.; TidAng, J.; Frangioudakis, G.; Beale, S.M.; Wright, L.E.; Zeng, X.Y.; Leslie, S.J.; Li, J.-Y.; Kraegen, E.W. Enhancement of muscle mitochondrial oxidative capacity and alterations in insulin action are lipid species dependent: Potent tissue-specific effects of medium-chain fatty acids. *Diabetes* **2009**, *58*, 2547–2554. [CrossRef] [PubMed]
83. Brandstetter, A.M.; Picard, B.; Geay, Y. Muscle fibre characteristics in four muscles of growing bulls: I. Postnatal differentiation. *Livest. Prod. Sci.* **1998**, *53*, 15–23. [CrossRef]
84. Huang, J.; Zhu, R.; Shi, D. The role of FATP1 in lipid accumulation: A review. *Mol. Cell. Biochem.* **2021**, *476*, 1897–1903. [CrossRef]
85. Jeong, J.; Kwon, E.G.; Im, S.K.; Seo, K.S.; Baik, M. Expression of fat deposition and fat removal genes is associated with intramuscular fat content in longissimus dorsi muscle of Korean cattle steers. *J. Anim. Sci.* **2012**, *90*, 2044–2053. [CrossRef]
86. Chen, X.; Luo, Y.; Wang, R.; Zhou, B.; Huang, Z.; Jia, G.; Zhao, H.; Liu, G. Effects of fatty acid transport protein 1 on proliferation and differentiation of porcine intramuscular preadipocytes. *Anim. Sci. J.* **2017**, *88*, 731–738. [CrossRef]
87. Qi, R.; Long, D.; Wang, J.; Wang, Q.; Huang, X.; Cao, C.; Gao, G.; Huang, J. MicroRNA-199a targets the fatty acid transport protein 1 gene and inhibits the adipogenic trans-differentiation of C2C12 myoblasts. *Cell. Physiol. Biochem.* **2016**, *39*, 1087–1097. [CrossRef] [PubMed]
88. Qiu, F.; Xie, L.; Ma, J.-E.; Luo, W.; Zhang, L.; Chao, Z.; Chen, S.; Nie, Q.; Lin, Z.; Zhang, X. Lower expression of SLC27A1 enhances intramuscular fat deposition in chicken via down-regulated fatty acid oxidation mediated by CPT1A. *Front. Physiol.* **2017**, *8*, 449. [CrossRef] [PubMed]
89. Guitart, M.; Osorio-Conles, Ó.; Pentinat, T.; Cebrià, J.; García-Villoria, J.; Sala, D.; Sebastián, D.; Zorzano, A.; Ribes, A.; Jiménez-Chillarón, J.C. Fatty acid transport protein 1 (FATP1) localizes in mitochondria in mouse skeletal muscle and regulates lipid and ketone body disposal. *PLoS ONE* **2014**, *9*, e98109. [CrossRef]
90. Lobo, S.; Wiczer, B.M.; Smith, A.J.; Hall, A.M.; Bernlohr, D.A. Fatty acid metabolism in adipocytes: Functional analysis of fatty acid transport proteins 1 and 4. *J. Lipid Res.* **2007**, *48*, 609–620. [CrossRef]
91. Martin, G.; Schoonjans, K.; Lefebvre, A.-M.; Staels, B.; Auwerx, J. Coordinate regulation of the expression of the fatty acid transport protein and acyl-CoA synthetase genes by PPARα and PPARγ activators. *J. Biol. Chem.* **1997**, *272*, 28210–28217. [CrossRef]
92. Qi, R.; Feng, M.; Tan, X.; Gan, L.; Yan, G.; Sun, C. FATP1 silence inhibits the differentiation and induces the apoptosis in chicken preadipocytes. *Mol. Biol. Rep.* **2013**, *40*, 2907–2914. [CrossRef]
93. Liu, X.; Li, S.; Wang, L.; Zhang, W.; Wang, Y.; Gui, L.; Zan, L.; Zhao, C. The effect of FATP1 on adipocyte differentiation in Qinchuan beef cattle. *Animals* **2021**, *11*, 2789. [CrossRef]
94. La Cour Poulsen, L.; Siersbæk, M.; Mandrup, S. PPARs: Fatty acid sensors controlling metabolism. *Semin. Cell. Dev. Biol.* **2012**, *23*, 631–639. [CrossRef] [PubMed]
95. Nakae, J.; Kitamura, T.; Kitamura, Y.; Biggs III, W.H.; Arden, K.C.; Accili, D. The forkhead transcription factor Foxo1 regulates adipocyte differentiation. *Dev. Cell* **2003**, *4*, 119–129. [CrossRef]
96. Ioannilli, L.; Ciccarone, F.; Ciriolo, M.R. Adipose tissue and FoxO1: Bridging physiology and mechanisms. *Cells* **2020**, *9*, 849. [CrossRef]
97. Dowell, P.; Otto, T.C.; Adi, S.; Lane, M.D. Convergence of peroxisome proliferator-activated receptor γ and Foxo1 signaling pathways. *J. Biol. Chem.* **2003**, *278*, 45485–45491. [CrossRef]
98. Attie, A.D. Insig: A significant integrator of nutrient and hormonal signals. *J. Clin. Investig.* **2004**, *113*, 1112–1114. [CrossRef] [PubMed]
99. Kast-Woelbern, H.R.; Dana, S.L.; Cesario, R.M.; Sun, L.; de Grandpre, L.Y.; Brooks, M.E.; Osburn, D.L.; Reifel-Miller, A.; Klausing, K.; Leibowitz, M.D. Rosiglitazone induction of Insig-1 in white adipose tissue reveals a novel interplay of peroxisome proliferator-activated receptor γ and sterol regulatory element-binding protein in the regulation of adipogenesis. *J. Biol. Chem.* **2004**, *279*, 23908–23915. [CrossRef] [PubMed]
100. Li, J.; Takaishi, K.; Cook, W.; McCorkle, S.K.; Unger, R.H. Insig-1 "brakes" lipogenesis in adipocytes and inhibits differentiation of preadipocytes. *Proc. Natl. Acad. Sci. USA* **2003**, *100*, 9476–9481. [CrossRef]
101. Teixeira, P.D.; Oliveira, D.M.; Chizzotti, M.L.; Chalfun-Junior, A.; Coelho, T.C.; Gionbelli, M.; Paiva, L.V.; Carvalho, J.R.R.; Ladeira, M.M. Subspecies and diet affect the expression of genes involved in lipid metabolism and chemical composition of muscle in beef cattle. *Meat Sci.* **2017**, *133*, 110–118. [CrossRef] [PubMed]
102. Reue, K. The lipin family: Mutations and metabolism. *Curr. Opin. Lipidol.* **2009**, *20*, 165. [CrossRef]
103. Zhang, P.; Reue, K. Lipin proteins and glycerolipid metabolism: Roles at the ER membrane and beyond. *Biochim. Biophys. Acta BBA Biomembr.* **2017**, *1859*, 1583–1595. [CrossRef]
104. Phan, J.; Reue, K. Lipin, a lipodystrophy and obesity gene. *Cell Metab.* **2005**, *1*, 73–83. [CrossRef]
105. Reue, K.; Donkor, J. Lipin: A determinant of adiposity, insulin sensitivity and energy balance. *Future Lipidol.* **2006**, *1*, 91–101. [CrossRef]

106. Péterfy, M.; Phan, J.; Xu, P.; Reue, K. Lipodystrophy in the fld mouse results from mutation of a new gene encoding a nuclear protein, lipin. *Nat. Genet.* **2001**, *27*, 121–124. [CrossRef]
107. Reue, K.; Xu, P.; Wang, X.-P.; Slavin, B.G. Adipose tissue deficiency, glucose intolerance, and increased atherosclerosis result from mutation in the mouse fatty liver dystrophy (fld) gene. *J. Lipid Res.* **2000**, *41*, 1067–1076. [CrossRef] [PubMed]

metabolites

Review

Enhancing Metabolism and Milk Production Performance in Periparturient Dairy Cattle through Rumen-Protected Methionine and Choline Supplementation

Bingjian Huang [1,2], Muhammad Zahoor Khan [1,3,*], Xiyan Kou [1], Yinghui Chen [1], Huili Liang [1], Qudrat Ullah [3], Nadar Khan [4], Adnan Khan [5], Wenqiong Chai [1] and Changfa Wang [1,*]

1 Liaocheng Research Institute of Donkey High-Efficiency Breeding and Ecological Feeding, Liaocheng University, Liaocheng 252000, China; 2210150218@stu.lcu.edu.cn (Y.C.)
2 College of Life Sciences, Liaocheng University, Liaocheng 252059, China
3 Faculty of Veterinary and Animal Sciences, University of Agriculture, Dera Ismail Khan 29220, Pakistan; qudrat.ullah@uad.edu.pk
4 Livestock and Dairy Development (Research) Department Khyber Pakhtunkhwa, Peshawar 25120, Pakistan
5 Genome Analysis Laboratory of the Ministry of Agriculture, Agricultural Genomics Institute at Shenzhen, Chinese Academy of Agricultural Sciences, Shenzhen 511464, China
* Correspondence: zahoorcau@cau.edu.cn (M.Z.K.); wangchangfa@lcu.edu.cn (C.W.)

Abstract: For dairy cattle to perform well throughout and following lactations, precise dietary control during the periparturient phase is crucial. The primary issues experienced by periparturient dairy cows include issues like decreased dry matter intake (DMI), a negative energy balance, higher levels of non-esterified fatty acids (NEFA), and the ensuing inferior milk output. Dairy cattle have always been fed a diet high in crude protein (CP) to produce the most milk possible. Despite the vital function that dairy cows play in the conversion of dietary CP into milk, a sizeable percentage of nitrogen is inevitably expelled, which raises serious environmental concerns. To reduce nitrogen emissions and their production, lactating dairy cows must receive less CP supplementation. Supplementing dairy cattle with rumen-protected methionine (RPM) and choline (RPC) has proven to be a successful method for improving their ability to use nitrogen, regulate their metabolism, and produce milk. The detrimental effects of low dietary protein consumption on the milk yield, protein yield, and dry matter intake may be mitigated by these nutritional treatments. In metabolic activities like the synthesis of sulfur-containing amino acids and methylation reactions, RPM and RPC are crucial players. Methionine, a limiting amino acid, affects the production of milk protein and the success of lactation in general. According to the existing data in the literature, methionine supplementation has a favorable impact on the pathways that produce milk. Similarly, choline is essential for DNA methylation, cell membrane stability, and lipid metabolism. Furthermore, RPC supplementation during the transition phase improves dry matter intake, postpartum milk yield, and fat-corrected milk (FCM) production. This review provides comprehensive insights into the roles of RPM and RPC in optimizing nitrogen utilization, metabolism, and enhancing milk production performance in periparturient dairy cattle, offering valuable strategies for sustainable dairy farming practices.

Keywords: dairy cattle; periparturient period; RPM; RPC; milk production; metabolism; nitrogen utilization

1. Introduction

Effective nutritional management during the periparturient period plays a crucial role in optimizing lactational performance among lactating dairy cows [1–3]. Previous research has emphasized that variations in dry matter intake (DMI) and concentrations of non-esterified fatty acids (NEFA) have a significant impact on both fertility and subsequent milk production outcomes [3,4]. The common practice of implementing a traditional dry period before parturition inadvertently results in reduced DMI, thereby limiting energy intake. Consequently, this leads to a state of negative energy balance, causing a surge in NEFA and β-hydroxybutyric acid (BHBA) levels. Elevated levels of NEFA and BHB can incite compromised immune function, particularly by dampening neutrophil activity in the lead-up to calving [5]. In addition, that elevation of NEFA and BHB regulate oxidative stress, which is followed by increased inflammatory changes and suppressed immunity [2,6]. The suppressed immunity, elevated oxidative stress and inflammatory changes due to negative-energy-balance-causing abnormal levels of NEFA and BHB are the key factors that expose dairy cattle to various diseases including mastitis [7,8], ketosis, retained placenta, and ketosis [9,10]. Furthermore, escalated NEFA and BHB levels with poor health also exert an adverse influence on lactational performance [7].

In the realm of dairy cattle nutrition, the significance of RPM and RPC extends beyond their roles in metabolism and milk production enhancement. These compounds have garnered substantial attention for their contributions to periparturient health disorders in dairy cattle. Research has elucidated the pivotal roles of RPM and RPC in alleviating oxidative stress, mitigating inflammatory changes, and bolstering immunity [2,11]. This body of evidence underscores their potential in promoting overall bovine health. Furthermore, several field experimental trials have consistently reported the clinical implications of RPM and RPC supplementation [12,13]. Notably, RPM has emerged as a crucial factor in controlling and reducing the risk of mastitis in dairy cattle during the periparturient period [2,12,13]. Additionally, Dai et al. [14] demonstrated through in vitro experimental procedures that methionine and arginine treatment effectively alleviated inflammatory changes and oxidative stress induced by lipopolysaccharide (LPS) in bovine mammary epithelial cells (BMECs). Moreover, it is well known that ruminants possess a limited capacity to secrete very low-density lipoproteins (VLDL) from the liver, which can hinder the export of hepatic triacylglycerols, ultimately resulting in hepatic lipidosis. However, the supplementation of 25.8 g/d of RPC has been found to significantly increase hepatic triacylglycerol secretion, thereby reducing the incidence of hepatic lipidosis in dairy cows [15,16]. Other studies have indicated that RPM can reduce the occurrence of subclinical hypocalcemia [17]. However, it should be noted that RPM supplementation did not appear to reduce the incidence of other health disorders, including retained placenta and ketosis, in dairy cattle.

To enhance milk production, dairy cattle during the periparturient period are provided with a high-protein diet. The dairy industry faces a well-established phenomenon where dairy cows convert 20 to 35% of dietary crude protein (CP) into milk, with the remaining nitrogen being excreted in manure, leading to detrimental environmental consequences [18–21]. In response to this issue, reducing CP supplementation has become a significant focus in the field of lactating ruminants in the United States, as CP serves as the primary nitrogen source [22]. The U.S. dairy sector is under increasing pressure to reduce emissions of nitrogen, phosphorus, and greenhouse gases [23]. Nitrogen, in particular, represents a primary pollutant originating from dairy farm operations, contributing to instances of nitrate contamination in groundwater, the eutrophication of surface water, and emissions of ammonia and nitrous oxide into the atmosphere [24,25].

Implementing the recommended reduction in mobilized protein supplementation, as advocated by the National Research Council (NRC) in 2001 [26], could lead to improved nitrogen utilization for milk production within dairy cattle [27]. However, empirical evidence has shown that dietary protein intake below the recommended threshold may compromise milk production, milk protein yield, and DMI [28,29]. In such cases, the incorporation of RPM [30–34], along with RPC supplementation [35,36], has emerged as an effective strategy to mitigate the adverse consequences resulting from inadequate dietary protein intake.

RPM and RPC have received significant attention due to their notable contributions in promoting milk production performance and optimizing nitrogen utilization within dairy cattle [36–41]. A series of studies have consistently confirmed the crucial role of amino acids in orchestrating the regulation of milk and its components in dairy cattle [42–46].

Consequently, a study has emphasized the indispensable role of RPM in milk production and metabolism, as evidenced by in vitro experiments using mammary epithelial cells. These studies highlight the relevance of concentrations of lysine (Lys) and RPM in the medium, not only in optimizing milk protein synthesis but also in influencing amino acid transport and signal transduction pathways that impact the expression of genes associated with milk protein synthesis [47]. Optimal levels of Lys and RPM, along with a supplemental Lys-to-RPM ratio approximating 3:1, have been identified as catalysts for eliciting the expression of genes instrumental in milk protein transcription and translation, ultimately resulting in the peak of casein production and cell proliferation rates [26,48]. RPM, in synergy with other amino acids, has similarly shown positive effects on the production performance and the metabolic aspects of dairy cattle. Importantly, a study has documented that milk yield [12], as well as energy-corrected milk (ECM), milk fat, protein, and lactose percentage levels, exhibited enhancements in cows supplemented with RPM and lysine compared to the control group [49,50]. Furthermore, the study ascertained that supplementation with RPM and RP-lys preserved the post-calving body condition score (BCS), enhanced DMI, and reduced the blood concentrations of BHBA [49,50].

The importance of RPM in milk production is accentuated by its status as the rate-limiting amino acid for lactating dairy cows, especially when fed protein supplements high in lysine but low in methionine, such as blood meal and soybean meal. To mitigate this constraint and augment the post-ruminal methionine supply, the provision of RPM has emerged as an effective strategy. A multitude of studies have outlined the advantages conferred by RPM supplementation, including the economization of dietary protein and the enhancement of overall lactational performance in dairy cows, particularly in diets formulated around standard North American ingredients [51,52]. Recent academic efforts have also focused on RPM supplementation in the diets of transition cows, revealing that achieving a Lys-to-RPM ratio approximating 2.9:1 as a fraction of metabolizable protein results in increased milk yield, partly attributable to increased DMI and the potential for the more effective utilization of bodily lipid reserves [52].

In addition, methionine, as the only amino acid containing sulfur, plays the role of a precursor for other sulfur-containing amino acids, namely cysteine, homocysteine (Hcy), and taurine, all of which play critical roles in various methylation reactions. The metabolic journey of methionine begins with its conversion into S-adenosylmethionine (SAMe), a key cofactor in methionine intermediate metabolism used in methylation reactions. SAMe serves as a methyl donor for amino acid residues within proteins, DNA, RNA, and small molecules, thereby influencing a wide range of biological processes [53]. Furthermore, methionine acts as a precursor for hydrogen sulfide, taurine, and glutathione, all of which have demonstrated their effectiveness in counteracting oxidative stress caused by various oxidants, thus protecting tissues from damage [54]. As a result, the dietary inclusion of rumen-protected methyl donors, such as RPM, has advantages in meeting the needs of cows during the peak of lactation, when the outflow of methylated compounds in milk is increased.

Extensive academic research has highlighted the positive correlation between RPC and improved production performance, while also providing improvement for metabolic challenges faced by dairy ruminants [55]. This suggests that RPC, similar to a vitamin-like substance, plays a crucial role in animal production, reproduction, and overall health [35]. Of particular note, an insightful meta-analysis study has illuminated the effectiveness of RPC, as evidenced by its ability to increase the production of ECM, elevate milk yields, and enrich the composition of milk constituents [56].

Given the significant importance attributed to RPM and RPC, this review aims to comprehensively explore their fundamental roles as indispensable nutritional strategies. The scope of this inquiry is directed towards enhancing metabolic processes, optimizing nitrogen utilization, and ultimately improving lactational performance within the domain of dairy cattle.

2. Materials and Methods

We conducted a comprehensive literature review encompassing studies that investigate the influence of RPM and RPC on the metabolic processes, nitrogen utilization, and milk production performance in periparturient dairy cattle. The identification of pertinent articles for this review was accomplished through rigorous search methodologies utilizing reputable academic databases, including PubMed, ScienceDirect, Web of Science, SpringerLink, Scopus, and Google Scholar.

The search strategy was thoroughly designed, employing key terms such as "milk production", "metabolism", "one-carbon metabolic pathway", "amino acid metabolism", "periparturient dairy cattle", "RPC", and "RPM" to retrieve the relevant literature. Our focus was primarily directed towards sourcing data published in the English language and featured in highly regarded peer-reviewed journals. Specifically, we prioritized articles published from the year 2000 onwards to ensure the inclusion of contemporary research findings.

To maintain the highest standards of academic rigor, we deliberately excluded content in the form of conference abstracts, books, book chapters, and unpublished findings from our review. This stringent selection process ensured that the information discussed in this review is grounded in the credible and peer-reviewed scientific literature, contributing to the overall reliability and integrity of our analysis.

3. Interplay of Methionine and Choline in One-Carbon Metabolism and Amino Acid Regulation

Methionine plays a crucial role in one-carbon metabolism [57], coordinating a complex series of biochemical processes that transfer one-carbon (methyl) groups in various forms [58–62]. This metabolic pathway is essential for critical cellular functions, including the synthesis of DNA and RNA, amino acid metabolism, and the production of S-adenosylmethionine (SAMe), a universal methyl donor crucial for methylation reactions throughout the body [63,64]. Methionine itself acts as the precursor to SAMe, a key molecule in this pathway, formed by combining methionine with adenosine triphosphate (ATP). SAMe, in turn, donates its methyl group to various substrates, participating in a wide range of biochemical reactions, including DNA and RNA methylation, protein methylation, and the synthesis of neurotransmitters, which are essential for mood regulation and brain function.

As SAMe exhausts its methyl group in methylation reactions [57], it transforms into S-adenosylhomosysteine (AdoHcy) [65]. This conversion involves further steps, including transferring a methyl group to Hcy to regenerate methionine. Methionine also initiates protein synthesis, a fundamental process responsible for constructing all body proteins, from enzymes to structural proteins, as well as those involved in various cellular functions. It actively participates in transmethylation reactions, influencing the synthesis of amino acids, neurotransmitters, DNA, RNA, and lipids. Furthermore, methionine contributes to its own regeneration from Hcy and the production of vital molecules such as creatine, choline, and epinephrine. It also aids in cysteine synthesis through transsulfuration, which is essential for the formation of protein disulfide bonds and the production of the antioxidant glutathione [66,67].

Choline plays a vital role in regenerating methionine from Hcy by generously donating methyl groups in one-carbon metabolism [68]. Additionally, choline is involved in creatine synthesis, which is crucial for muscle energy metabolism, and serves as a precursor for choline-containing amino acids such as betaine [69], and sarcosine, contributing to Hcy conversion into methionine and methyl group detoxification. In conclusion, methionine and choline intricately interconnect in amino acid metabolism [70]. Methionine supports protein synthesis, transmethylation reactions, and cysteine synthesis, while choline contributes to methionine regeneration [71], creatine synthesis, and the production of choline-containing amino acids. Together, these processes ensure the availability of amino acids, crucial for protein production, cellular functions, and overall health in dairy cattle and other organisms.

From the above discussion, it can be concluded that methionine and choline have distinct but interconnected roles in amino acid metabolism. Methionine serves as an essential amino acid required for protein synthesis and participates in transmethylation reactions and cysteine synthesis. Choline, on the other hand, is crucial for regenerating methionine from Hcy, contributing to the one-carbon metabolism pathway, and plays a role in the synthesis of creatine and choline-containing amino acids. These processes collectively ensure the availability of amino acids for protein production, cellular functions, and overall health in dairy cattle and other animals.

4. The Role of RPM and RPC in Metabolism and Milk Production Performance of Dairy Cattle

4.1. RPM Role in Metabolism and Nitrogen Utilization of Dairy Cattle

Methionine, an indispensable amino acid, plays a crucial role in shaping the physiological development, metabolism, and growth of mammals [72–74]. Seymour particularly emphasizes the importance of this amino acid in metabolic processes, especially as a precursor to essential compounds required for vital physiological functions [75]. Methionine serves as a fundamental building block for compounds such as succinyl-CoA, Hcy, cysteine, creatine, and carnitine. Furthermore, its active involvement in the biosynthesis of SAMe is essential for polyamine, creatine, and phosphatidylcholine metabolism [76].

The role of methionine extends far beyond its basic building-block function. It actively participates in cellular methylation reactions and sulfur recycling processes, with the ability to undergo enzymatic conversion to L-methionine sulfoxide [76,77]. The resulting product of this conversion, cysteine, plays a critical role in fundamental cellular functions, including protein translation, glutathione synthesis, and taurine production [78,79]. The consequences of insufficient methionine levels are significant, leading to detrimental effects such as small intestine atrophy, suppressed epithelial growth in neonatal animals, reduced goblet cells, and diminished glutathione content within the small intestine [80].

Within the gastrointestinal tract (GIT), approximately 20% of dietary methionine is absorbed and utilized [81]. Research by Shoveller et al. [82] indicates that neonatal piglets' parenteral methionine requirement approximates 69% of the enteral requirement. Additionally, methionine catabolism has been found to be more pronounced in extracellular cells outside pig enterocytes, particularly in the portal-drained viscera and intestinal mucosa, as highlighted by Blachier et al. [83]. The dynamic interaction of methionine with choline, as emphasized by Swain and Johri [84], underscores its significance in antibody (IgG) production. Conversely, methionine deficiency leads to a decrease in relative lymphoid organ weight, ultimately impacting overall growth performance [85]. The balance is delicate, as excessive methionine intake has been associated with growth depression [86].

A comprehensive exploration of the role of amino acids in nitrogen management has been a topic of close scrutiny [32,87–90]. Recent research has provided significant insights. For example, our own study found that cows in the RPM group had lower blood-urea nitrogen concentrations compared to the control group, along with higher levels of rumen microbial CP [91]. Consistently, academic studies have supported the positive impact of supplementing RP-Lys and RPM, combined with corn grain and soybean meal, in reducing urinary urea nitrogen excretion [92]. Further research by Ding et al. [93] confirmed the enhanced nitrogen utilization efficiency in dairy cattle through the infusion of arginine and RPM. Empirical evidence further emphasizes the beneficial effects of methionine supplementation in addressing environmental issues arising from excessive nitrogen emissions [94]. Additionally, it has been documented that a balanced combination of RP-lysine, RPM, and threonine can improve nitrogen absorption in dairy heifers, thereby contributing to reduced environmental nitrogen levels. Furthermore, concurrent supplementation of RPM and RP Lys leads to metabolic improvements and decreased nitrogen excretion [95]. These collective findings underscore the crucial role of amino acid supplementation as a promising strategy to address nitrogen-related environmental challenges and enhance nutrient utilization in livestock. Such strategies have the potential to significantly promote sustainable agricultural practices while simultaneously enhancing animal health and overall productivity.

In a study by Sobhanirad et al., the comparison of organic and inorganic zinc supplements' effects on milk production and composition among lactating dairy cows revealed interesting insights. Although the source of zinc did not significantly affect the milk and FCM yield, the basal diet supplemented with zinc methionine (ZnM) showed potential benefits. This supplementation resulted in a higher milk and FCM yield, with a lower somatic cell count, compared to controls [96]. Consistent with these findings, an earlier study demonstrated improved lactational performance and reduced somatic cell counts in cows fed ZnM [97]. These results highlight the potential advantages of ZnM supplementation in enhancing dairy cow performance and milk quality.

4.2. Rumen-Protected Methionine's Role in Ruminants' Milk Production

Dairy animals during parturition are prone to experiencing a negative energy and protein imbalance because their nutrient intake is insufficient to meet the demands of milk production [6,8,98]. The importance of RPM in milk production has garnered significant attention in dairy cattle research [99–103]. Therefore, increasing milk production requires the strategic implementation of high-protein diet supplementation for dairy cattle. This approach is supported by substantial evidence, as there is a well-documented association between RP Lys, RPM, and threonine concentrations with the growth, physiology, and reproductive performance of calves [104,105]. Accordingly, Lee et al. emphasized that providing RP-Lys, RPM, and histidine in combination significantly increased milk protein yield in dairy cows fed a diet deficient in metabolizable protein [106]. Further evidence suggests that supplementing cows with 10 g/day of RPM, along with a concentrated diet containing corn grain and soybean meal, significantly increased milk production in dairy cattle [107]. This finding is consistent with the study of Carder and Weiss [108], who demonstrated that supplementation with RPM and Lys resulted in sustained increases in milk energy and milk yield. Consistently, the continuous supplementation of RPM has been linked to increased milk yield and milk protein content in periparturient dairy cows [109,110]. Studies conducted during the pre- and postpartum periods consistently showed that RPM supplementation can increase both milk yield and milk protein content in dairy cattle [13,17,111]. Giallongo et al. further supported these findings by demonstrating that combining RPM increased overall milk performance, including milk yield and milk components, in Holstein cows [33]. Furthermore, RPM supplementation has been shown to effectively enhance milk performance and address the metabolizable-protein deficiency gap in dairy cattle [106,112]. Even in the context of a diet deficient in metabolizable protein, milk production remained unaffected in dairy cattle supplemented with RPM [87]. Recent research by Park et al. reaffirms the positive impact of supplementing cows with 10 g/day of methionine, in conjunction with corn grain and soybean meal, resulting in a substantial enhancement in milk production. Additionally, studies have experimentally demonstrated the benefits of supplementing corn silage and corn milling products with RPM, yielding notable improvements in milk production among dairy cattle [113]. Furthermore, the genetic factors underlying the effects of RPM supplementation on milk production performance have been the focus of extensive investigation.

From the above discussion, it is clear that RPM supplementation plays a crucial role in addressing the challenges of negative energy and protein balance during parturition. An increasing body of research consistently demonstrates that careful RPM supplementation, along with strategic dietary adjustments, holds promise in significantly increasing milk yield and enhancing overall milk quality in dairy cattle.

4.3. Molecular Mechanisms Unveiling the Influence of Methionine Supplementation on Dairy Cattle Milk Production Performance

A comprehensive investigation into the genetic basis of the impact on milk production in dairy cattle due to RPM supplementation has revealed intricate molecular pathways and regulatory elements. Numerous in vitro experimental studies have diligently elucidated the regulatory functions and molecular mechanisms of cAMP response element-binding protein-regulated transcription coactivator 2 (CRTC2) in methionine-induced milk fat synthesis [45]. These studies emphasize the critical role of CRTC2 as a transcription coactivator within the methionine-induced milk fat synthesis pathway mediated by the mammalian target of rapamycin (mTOR) in BMECs.

An independent study has demonstrated that methionine supplementation has a significant influence on the expression of purine-rich element-binding protein B (PURB), a key regulator of gene transcription and cellular physiology [46]. The upregulation of PURB, in conjunction with methionine treatment, has led to increased milk protein and fat synthesis, accompanied by elevated expressions of mTOR and sterol response element-binding protein (SREBP)-1c within BMECs. Interestingly, counteractive effects were observed when PURB expression was reduced. Furthermore, an important discovery has emerged regarding the positive modulation of U2 snRNP auxiliary factor 65 kDa (U2AF65) by methionine, further enhancing milk synthesis and cell proliferation within BMECs through the mTOR-SREBP-1c signaling pathway [114].

Annexin A2 (AnxA2), renowned for its diverse roles encompassing growth, development, and metabolism, assumes a significant role in milk synthesis and cell proliferation. Methionine treatment has exhibited a positive influence on phosphatidylinositol 3-phosphate (PIP3) levels, mTOR phosphorylation, and protein levels of SREBP-1c and Cyclin D1. In this complex process, AnxA2 emerges as a critical regulator through the phosphatidylinositol-3-kinase (PI3K)-mTOR-SREBP-1c/Cyclin D1 signaling pathway [115]. Furthermore, the synergistic overexpression of glucose-regulated protein 78 (GRP78) alongside methionine treatment has yielded notable stimulatory effects on milk protein and milk fat synthesis, augmented cell proliferation, and an affirmative modulation of mTOR phosphorylation. This coalescence has also heightened protein levels of Cyclin D1 and SREBP-1c. Notably, the ablation of GRP78 through siRNA transfection has manifested contrasting outcomes. Intriguingly, the predominant cytoplasmic localization of GRP78 in bovine mammary epithelial cells has been observed. Moreover, its protein expression has been significantly enhanced following stimulation with methionine, leucine, estrogen, and prolactin [116].

Implicating the nuclear factor of κB (NFκB) family, renowned for its roles in gene expression regulation, unveils its involvement in milk synthesis regulation. NFκB1 has been identified as a governing factor for various genes, including SREBP-1c, and β4-galactosyltransferase-T2 (β4Gal-T2), consequently impacting the milk biosynthesis process [117]. Methionine treatment has been shown to enhance NFκB1's binding to gene promoters of mTOR, SREBP-1c, and β4Gal-T2 in BMECs, illuminating milk biosynthesis to be facilitated through the PI3K pathway rather than the mTOR signaling pathway. Further insights have emerged from studies delving into the intricate regulatory network, indicating that methionine treatment leads to the suppression of DEAD-box helicase 6 (DDX6) expression. As a pivotal member of the RNA helicase family governing mRNA storage and translation regulation, decreased DDX6 expression adversely affects milk synthesis by hampering the effects of p-mTOR, SREBP-1c, and Cyclin D1 within BMECs [118]. Furthermore, gene functional analyses have corroborated the positive regulatory impact of methionine on SREBP-1c gene expression, thereby promoting milk fat synthesis through epidermal-type fatty acid binding protein-5 (*FABP-5*) [119]. Intriguingly, the heterodimeric amion acid taste receptor (TAS1R1/TAS1R3) constellation emerges as a sensor for extracellular methionine in BMECs. Activation of this receptor tandem triggers mTOR signaling, possibly via intracellular calcium-concentration elevation, and is implicated in the mediation of methionine and valine-induced changes in β-casein (CSN2) mRNA abundance [120].

A comprehensive overview of the genetic responses elicited by methionine in the regulation of milk synthesis, has been presented in Table 1. These findings collectively illuminate the intricate genetic mechanisms by which rumen-protected methionine supplementation exerts its influential role in enhancing milk production performance in dairy cattle.

Table 1. Summary of studies investigating the influence of RPM on metabolism and milk production performance of dairy cattle.

Amino Acid Supplementation	Main Outcomes	Species	Author
RPM and RPC	✓ Enhanced metabolism and milk production performance	Dairy cattle	[121]
RPM	✓ Improved metabolism, enhanced milk production	Dairy cattle	[122]
RPM	✓ The RPM supplementation promoted higher milk yield, ECM yield, milk protein yield, milk protein, casein and milk fat yield	Dairy cattle	[123]
RPM	✓ Increased milk yield and milk components yields and metabolism	Dairy cattle	[112]
ZnM	✓ Improved lactational performance ✓ Facilitated metabolic processes, enhanced nitrogen utilization, and decreased environmental pollution	Dairy cattle	[98,124]
RPM and RPC	✓ Enhanced milk production performance, increased postpartum dry DMI, regulated hepatic lipid metabolism, and improved immunity	Dairy cattle	[125]
Hydroxyselenomethionine	✓ Enhanced lactation performance, and milk Se concentrations in early-lactating dairy cows	Dairy cattle	[126,127]
N-acetyl-l-methionine supplementation (NALM)	✓ Enhanced milk production ✓ Increased concentrations of total protein and globulin in plasma ✓ Improved nitrogen absorption and alternative reduction in environmental pollution	Dairy cattle	[128,129]
NALM s	✓ Facilitated metabolism with improvement in milk production performance	Dairy cattle	[130,131]
RPM	✓ Enhanced metabolism of carnitine and enhanced β-oxidation of fatty acids, and improved cholesterol metabolism followed by Lipoprotein metabolism ✓ Promoted one-carbon-metabolism cystathionine and beta-synthase-activity cystathionine, followed by enhancement of antioxidant synthesis	Dairy cattle	[132]
RPM	✓ Enhanced digestion and milk production ✓ Improved utilization of nitrogen absorption	Dairy cattle	[133]
RPM and RPC	✓ Enhanced metabolism and milk production, and decreased incidence of metabolic issue	Dairy cattle	[134]
16% CP and 25 g/head/day RPM	✓ Enhanced milk production phenotypes (milk casein, milk yield and milk fat yield)	Dairy cattle	[135]
RPM and RP Lys	✓ Promoted yield of milk, fat, protein, and lactose ✓ RPM and RPL did not change milk production phenotypic traits in adequate amino acids diet	Dairy cattle	[136]
RPM+RP Lys	✓ Milk production performance including milk yield and the content of fat and protein were significantly enhanced ✓ Metabolism was improved and it prevented body weight loss	Dairy cattle	[137]

Table 1. *Cont.*

Amino Acid Supplementation	Main Outcomes	Species	Author
RPM+fava bean	✓ Improved DMI followed by enhancement of milk protein yield ✓ Nitrogen-use efficiency for milk production tended to increase by decreasing nitrogen excretion ✓ Elevated plasma methionine	Dairy cattle	[138]
RPM and RP Lys for 21 days	✓ Increased DMI and increased milk yields, and yield of milk protein, solids, lactose, and milk lactose concentration ✓ Promoted milk Se ng/mL, milk N g/kg and milk Se:N, ng/kg	Dairy cattle	[139]
RPM	✓ Significantly enhanced the milk yield, milk protein percentages and milk fat percentages	Dairy cattle	[140]
RPM and RPC	✓ Enhanced metabolism and amino acid profile in plasma followed by improved lactational performances	Dairy cattle	[121]
RPM	✓ Induced ARID1A degradation to promote mTOR expression and milk synthesis in mammary epithelial cells	BMECs	[41]
RPM	✓ Increased the expression of PURB to stimulate mTOR and SREBP-1c, followed by increased milk protein and fat synthesis	BMECs	[46]
RPM	1. Enhanced the expression of Twinfilin1 (TWF1) which, by using the mTOR pathway, enhances the milk bio-synthesis and cell proliferation	BMECs	[141]
Methionyl-methionine dipeptide	✓ Regulated the synthesis of β-CSN in BMECs. ✓ Increased the mRNA abundance of Janus kinase 2 (JAK2) and signal transducer and activator of transcription 5 (STAT5), leading to the enhanced phosphorylation of JAK2, STAT5, mTOR, p70 ribosomal S6 kinase 1, and eukaryotic initiation factor 4E binding protein 1. ✓ Inhibition of both JAK2 and mTOR significantly reduced the Met-Met-induced increase in cell viability and β-CSN synthesis in BMECs.	BMECs	[142,143]
RPM	✓ β-CSN expression was enhanced followed increased level of mTOR and SREBP-1c, resulting in milk and milk fat synthesis	BMECs	[144]
RPM	✓ Biosynthesis of milk β-CSN via regulating mTORC2-protein kinase B (AKT) signaling in methionine (Met)-induced L-type amino acid transporter 1 (LAT1) ✓ Upregulated milk protein synthesis	BMECs	[145]
RPM	✓ Activates the TAS1R1/TAS1R3 receptor, which appears to function as a sensor for extracellular methionine in bovine mammary cells, leading to the activation of mTOR signaling, possibly through the elevation of intracellular calcium concentration and mediating the changes in CSN2 mRNA abundance ✓ Also promoted the activation of CSN1S1 and eukaryotic elongation factor (eEF) 2 to enhance milk biosynthesis	BMECs	[120,146–148]
RPM	✓ Enhanced mTOR phosphorylation via Wnt-induced secreted protein 3 (WISP3), which is followed by improved milk fat synthesis	BMECs	[149]

4.4. Rumen-Protected Choline Role in Metabolism of Dairy Cattle

Choline, an indispensable nutrient, plays a central role in the field of metabolism. It is embedded in the composition of lipid-soluble metabolites, especially phosphatidylcholine, lysophosphatidylcholine, free choline and sphingomyelin, all of which are essential components of cell membranes. These compounds play critical roles in cell signaling and lipid metabolism, and their profound importance has been clearly elucidated by the scientific community [150]. The indispensable nature of choline and its metabolites underscores their important roles in maintaining structural integrity and promoting signaling functions within cell membranes. Furthermore, choline actively participates in the synthesis of acetylcholine, a highly important neurotransmitter, and also contributes to methylation processes through the efficient transfer of methyl groups, facilitated by the intermediary betaine metabolite, leading to the synthesis of the crucial SAMe pathway [151].

The prominence of choline metabolites in the biological development of mammals is particularly highlighted during the transition period. Supporting this, observations have confirmed the potential for choline supplementation during this critical phase to enhance hepatic lipid metabolism in cells, a finding supported by various studies [152–154]. Of note, among these choline metabolites, betaine emerges as a key facilitator in the synthesis of creatine, carnitine, and methionine. Its role as a methyl group donor, which traverses the transmethylation pathway, has been identified as instrumental in this process [155]. A spectrum of dietary sources boasts substantial phosphatidylcholine content, typically around 13% by weight [156]. Interestingly, the inclusion of choline in the diet significantly influences the remethylation activity of betaine. While the reversal of betaine into choline through a reverse reaction is not feasible, betaine assumes a supporting role under circumstances where choline availability is constrained, aiding the remethylation reaction for the synthesis of Hcy [157]. This dynamic modulation involving betaine orchestrates methylation through the Hcy methyltransferase reaction, thereby exerting regulatory control over adoHcy and S-adenosyl methionine levels within cells. This, in turn, leads to improvements in the epigenetic mechanism and DNA methylation within the cellular environment. Moreover, the presence of choline and betaine within the diet, alongside other methyl donor groups, exerts a profound influence on methylation reactions [138]. Of particular significance, the collaborative contributions of methyl-tetrahydrofolate and betaine underscore their involvement in the biosynthesis of methylation from Hcy to methionine [158].

During the pivotal transition period, dairy cows undergo significant physiological changes. These changes include a reduction in dry matter intake, an imbalance between negative energy and metabolizable protein, and an increased demand for high-quality nutrients to support both fetal growth and lactation performance. This juncture is accompanied by the emergence of potential challenges such as ketosis, hypocalcemia, clinical mastitis, and indications of fatty liver [159–161]. The liver functionality index (LFI), serving as a reflection of transition cow metabolic health, captures shifts in biomarkers linked to liver plasma protein synthesis (albumin), lipoprotein synthesis (cholesterol), and heme catabolism (bilirubin). Importantly, research has illuminated the connection between choline-fed cows and elevated LFI values, translating to improved dry matter intake, milk yield, milk fat yield, milk protein yield, and a reduced vulnerability to metabolic ailments [66].

During this complex transition phase, the interplay of stress and hormonal dynamics leads to an increase in NEFA in circulation, serving as a compensatory mechanism to counter the negative energy balance in transitioning cows. The liver coordinates the oxidation of NEFA to generate energy, with some of it redirected towards the production of TG, subsequently mobilized as VLDL [162]. However, it is important to note that a portion of NEFA undergoes partial oxidation in the liver, resulting in the accumulation of ketone bodies in the bloodstream, hindering the transport of triglycerides through VLDL. Additionally, in the pre- or early lactation phase for cows, there may be an elevation in the ratio of lipid peroxidation [163], a reduction in serum α-tocopherol, and increased levels of oxidative stress. This collective situation has the potential to negatively impact the health and productive performance of dairy cows [164]. Notably, RPC supplementation has been confirmed to help lower blood BHBA levels in periparturient dairy cattle [165]. Existing evidence highlights the role of RPC in improving hepatic TG levels, thereby supporting optimal production performance during the lactational period [37,166]. Introducing dietary RPC supplementation yields significant results, with a noticeable increase in colostrum production among Holstein cows [167,168]. This increase is particularly pronounced among cows in their second parity. However, it is worth considering that the demand for choline could potentially be even more pronounced among cows in their third or higher parities. Furthermore, strategic choline supplementation has been found to translate into higher concentrations of phosphocholine within colostrum. Notably, RPC supplementation has been associated with an elevation in trimethylamine N-oxide concentrations within colostrum [167,168]. A compelling study conducted by Amrutkar et al. demonstrated that RPC supplementation, administered at a dose of 54 g/day for 40 days pre-calving and 120 days post-calving, significantly increased the levels of phosphatidylcholine and vitamin E in dairy cattle [169].

4.5. RPC Role in Milk Production of Dairy Cattle

The pivotal role of RPC in enhancing the lactational performance of dairy cattle has been extensively studied and is concisely presented in Table 2. The use of RPC in peripartum dairy cattle has garnered significant attention due to its consistent ability to increase either milk yield or fat-corrected milk production [170]. This phenomenon has received ample confirmation in the scientific literature, with a consensus gradually emerging that the inclusion of 12-20 g per day of RPC in the diet of cows results in optimal improvements in dairy cow production performance [56,171].

An in-depth meta-analysis conducted by Arshad et al. [56] has drawn widespread attention, focusing specifically on RPC supplementation. This meticulous effort compiled data from 23 distinct experiments, encompassing 74 treatment means and a robust sample of 1,938 cows. The results derived from this meta-analysis are striking, revealing a statistically significant increase in both pre- and postpartum daily dry matter intake by 0.28 kg per day and 0.47 kg per day, respectively, with significant differences. Furthermore, this analysis unveiled a noticeable rise in ECM production, resulting in a weighted mean average increase of 1.61 kg per day. Additionally, distinct improvements were observed in fat and protein yield, with increases of 0.08 kg per day and 0.06 kg per day, respectively [56]. Consistently, Humer et al. [55] conducted an exhaustive review of data from 27 separate studies, shedding light on a significant increase in postpartum dry matter intake, from an average of 19.1 to 19.9 kg per day ($p < 0.01$), along with an increase in milk yield from an average of 31.8 to 32.9 kg per day ($p = 0.03$) following RPC supplementation. Moreover, a statistically significant increase in milk fat and protein yield was observed ($p \leq 0.05$), while the compositional attributes of milk remained unaffected [55].

During the early stages of lactation, there is often a negative nutrient balance, leading to a decrease in choline metabolite concentrations in bovine plasma [154,172]. However, it has been discovered that RPC supplementation can reverse this trend, resulting in increased plasma concentrations of choline metabolites in prepartum cows, thereby improving their choline status [166]. Increasing choline status in the prepartum phase offers a promising approach to reduce the risk of fatty liver in cows [166,173] and positively impact postpartum health and lactation performance [174]. The research landscape in this domain involves an extensive exploration of the effects of RPC supplementation in various formulations, dosages, and dietary contexts on postpartum performance. Generally, including RPC in the diets of transition cows has yielded favorable results, demonstrating enhancements in milk, fat, and ECM yields [175,176]. However, it is essential to acknowledge the variability in responses across studies. Notably, the response to RPC supplementation has not been consistent in all investigations [121,177]. Additionally, the timing and duration of RPC feeding have been identified as pivotal factors affecting the performance of lactating Holstein cows, as elucidated by another study [38].

Table 2. Summary of studies associated with RPC impact on milk production and metabolism in dairy cattle.

RPC Treatment	Main Outcomes	Author
RPC at 15 g/day/head.	✓ RPC supplementation plays a substantial role as a lipotropic agent in the context of dairy ruminants. It operates by mitigating the accumulation of excess fat within hepatic cells, thereby promoting liver health. ✓ An observable outcome of RPC supplementation is a notable increase in various vital milk components. This includes a significant elevation in milk yield, as well as heightened levels of fat, lactose, solid-not-fat (SNF), total solids, and the protein yield. ✓ This cumulative effect of RPC supplementation contributes to the enhancement of milk quality, composition, and overall productivity within the dairy ruminant sector.	[171]
RPC at 54 g for 40 days before and 120 days after calving.	✓ Improved milk yield and milk component (fat, protein and lactose) yield. ✓ Suppressed the level of triglycerides and very low-density lipoproteins. ✓ Enhanced phosphatidylcholine and vitamin E levels.	[169]
Supplementing 12.9 g/d of choline ion.	✓ Supplementing 12.9 g/d of choline ions led to a 2.13 kg/d increase in ECM. ✓ This may be because choline supplementation prevents metabolic diseases and stress in lactating dairy cows.	[56]
The RPC supplementation was initiated during the final month of the transitional period and two months postparturient at 50 g/day/head.	✓ Supplementation with RPC resulted in a 2 kg/day/head increase in milk production, as indicated by the obtained results. ✓ Due to higher digestibility and increased total VFA concentration, and the prevention of metabolic disorders such as ketosis and fatty liver syndrome, this increase promotes milk fat synthesis by facilitating phospholipid synthesis, lipid absorption, and transport to the mammary gland.	[178]
The RPC product supplemented (60 g/d ReaShure, Balchem Corp., New Hampton, NY, USA) contained 28.8% choline chloride as per manufacture information, which supplied supplemented cows a daily dose of 12.9 g of choline ion.	✓ Modification of RPC increased the milk yield per animal by 2 L per day. Increased choline availability during early lactation may have stimulated the enzyme to increase mitosis in CT mammary cells. ✓ In addition, choline may exert some endocrine control over the mammary organ and influence nutrient partition toward milk synthesis via growth hormone increases. ✓ These modifications increased lipolysis and ketogenesis, which could provide nutrients for milk production. ✓ Moreover, RPC treatment decreased the liver triacylglycerol concentration of plasma in dairy cattle.	[38]
RPC supplementation.	✓ Milk production enhanced, decreased liver triacylglycerol concentration of plasma.	[179]
RPC supplementation.	✓ Increased yields of milk by 1.8 kg/d, fat by 0.08 kg/d, lactose by 0.08 kg/d, true protein by 0.04 kg/d, ECM by 1.9 kg/d, and fat-corrected milk by 2.1 kg/d without dry matter intake being affected.	[37]
200 g RPC after calving for 60 days of lactation, which provided 50 g per day choline.	✓ Throughout the research, RPC supplementation had a significant impact on milk production. During the first 60 days of lactation, the milk yield increased by 4.41 kg/day, and as a result, 4% FCM increased by 2.5 kg/day on average.	[180]
22 g choline ion from an established product (prepartum: 0.10 ± 0.004 choline ion, %DM.	✓ In general, peripartum RPC supplementation at the suggested dose tended to increase ECM yield after the supplementation period. ✓ This is because the effects of RPC on metabolic and inflammatory biomarkers support the potential for RPC supplementation to influence the metabolism and health of transition cows and may contribute to the observed production gains.	[181]

Table 2. *Cont.*

RPC Treatment	Main Outcomes	Author
RPC at 15 to 50 g/day/head.	✓ The RPC effect increased milk yield by an average of 1.05 kg per day. Additionally, the yields of milk protein and milk lipids were increased. This may be due to the prevention of metabolic condition, stress and dry matter intake. ✓ The RPC intake improved energy and nutrient intake is anticipated to have contributed to the reduction in NEB in early-lactating cows. This ameliorating effect is anticipated to manifest as either an enhanced NEB, i.e., less fat mobilization, reflected in less BW loss, lower circulating NEFA and BHBA, and higher glucose and cholesterol, or increased milk yields.	[55]
The RPC supplemented at 40 g/d/head.	✓ Increasing the intestinal supply of choline improved milk production in lactating dairy cows by approximately 7% over controls. ✓ Supplementation of RPC increased the milk production for the following reasons: higher digestibility and increased total VFA concentration, decreased NH3-N, and the prevention of metabolic disorders such as ketosis and fatty liver syndrome.	[35]
RPC at 54 g for 90 days (pre- and postcalving).	✓ In comparison to the control group, the prill fat alone- and prill fat + rumen-protected choline-supplemented groups produced 12.10 and 21.76% more milk (kg/d) on average. ✓ The higher milk production in supplemented groups was attributed to higher TDN, which increased the energy density of the ration and reduced the deleterious effect of a negative energy balance, as evidenced by lower blood NEFA levels.	[182]
RPC at 10 g 20 days pre-calving and 64 days post-calving.	✓ RPC supplementation increased milk yield up to 7.87% higher, compared to the without-RPC group during 21–64 DIM. ✓ This is due to the fact RPC is used for liver protection, enhances fat metabolism in the liver, and reduces triglyceride levels simultaneously. As a consequence, the liver generates more glucose, which is utilized for increased milk synthesis.	[183]
RPC at 30 g for 8 weeks after calving.	✓ The incorporation of RPC into the diet yielded noteworthy improvements in actual milk production, alongside a 4% increase in the yield of FCM. ✓ When compared to the unsupplemented RPC regimen, the addition of 30 g of RPC led to a substantial rise in actual milk yield, amounting to 2.24 kg per head per day, reflecting a remarkable increase of 14.85%. ✓ Improved nutrient digestibility and elevated concentrations of total volatile fatty acids (TVFA), coupled with reduced ammonia–nitrogen (NH3-N) concentration within the rumen of animals receiving RPC supplementation. ✓ RPC supplementation was associated with enhanced milk yield in high-producing dairy cows during the initial 60 days of lactation, potentially by augmenting lipid export from liver metabolism. ✓ It is recommended to employ RPC supplementation for high-yielding dairy herds, particularly in the peri-parturient period and especially when cows exhibit excessive body condition. ✓ Feeding RPC induces an increase in choline concentration and milk yield, indicative of augmented choline availability among supplemented cows. ✓ The substantial elevation in milk production is a direct outcome of metabolic adaptations induced by RPC supplementation, underlining the multifaceted impact of RPC on dairy cow physiology and milk performance.	[184]
RPC and vitamin B supplementation.	✓ Decreased liver fat and enhanced milk production.	[185]
60 g/d of RPC (13.0 g/d of choline ion).	✓ RPC increased the overall milk yield of primiparous cows by 3.1 kg/day. This may be due to a decrease in the negative energy balance and metabolic diseases due to choline supplementation.	[186]

Table 2. *Cont.*

RPC Treatment	Main Outcomes	Author
Supplemented 60 g of RPC/d (providing 15 g choline chloride) during the first 10 weeks of lactation.	✓ In the initial 10 weeks of lactation, dairy cows fed a total mixed ration (TMR) experienced enhanced milk production outcomes through the supplementation of 60 g of RPC (rumen-protected choline) per day. ✓ This supplementation regimen provided 15 g of choline chloride to the cows' diet. ✓ As a result of this supplementation strategy, the yield of FCM with a fat content of 4% (4% FCM) increased significantly by 3.28 kg per day. ✓ Additionally, there were notable improvements in the yields of both fat and protein from the milk produced by the supplemented cows.	[187]
RPC at 30 g/day per cow for the first 16 d of lactation.	✓ Cows supplemented with RPC at a rate of 30 g per head per day exhibited an increase in milk production of 3.8 kg per day, as compared to control cows, within the initial 16 days of lactation. ✓ RPC supplementation might have contributed to the reinforcement of the cows' antioxidant defense mechanisms, resulting in a reduction in oxidative stress within their systems. ✓ This supplementation regimen could have led to an elevation in the concentrations of essential nutrients such as vitamin E and methionine in the cows' physiology. ✓ Vitamin E, known for its potent antioxidant properties, could have played a role in minimizing oxidative stress, promoting better overall health, and potentially leading to increased milk production. ✓ Increased methionine concentrations, facilitated by RPC supplementation, could have positively influenced protein synthesis and metabolic processes associated with milk production.	[188]
RPC supplementation.	✓ Enhanced milk production and improved metabolism.	[189]
RPC was supplemented at 17.3 g/d/per cow from 21 days before the calving date to 21 days after.	✓ The ingestion of RPC exhibited a propensity to elevate milk production by approximately 2.2 kg/d per cow, even in the absence of discernible enhancements in liver function. ✓ This notable enhancement in milk production could potentially be attributed to amplified DMI and heightened energy mobilization. ✓ Cows subjected to RPC demonstrated a heightened intake of feed dry matter by 0.6 kg/d, indicating a potential avenue for the observed milk production increase. ✓ Cows consuming RPC showcased the elevated oxidation of fatty acids for energy during the initial stages of lactation. This energy diversion potentially conserved glucose for augmented lactose synthesis, thereby contributing to increased milk production.	[176]
RPC at 60 g/d.	✓ RPC is postulated to hold a higher likelihood of conferring favorable outcomes in overfed calves, as their susceptibility to hepatic triglycerides accumulation is elevated. ✓ The introduction of supplemental RPC resulted in heightened milk production among cows possessing a BCS of 4 at the commencement of the lactation phase. This increase is predominantly attributed to the augmented DMI. ✓ Based on these empirical findings, a hypothesis was formulated suggesting that supplementing herds where more than 20% of cows entered the transition period with a BCS of 4, with RPC, could potentially yield a positive impact on milk production.	[15,121]

Table 2. *Cont.*

RPC Treatment	Main Outcomes	Author
Dose of RPC was from 15 to 50 g.	✓ RPC supplementation exhibited a significant enhancement in milk yield during the commencement of lactation. ✓ RPC serves as both a precursor for acetylcholine and a vital source of methyl groups, facilitating the remethylation of Hcy into methionine. ✓ RPC supplementation potentially exerts a sparing influence on methionine, a pivotal amino acid in milk protein synthesis for lactating dairy cows, possibly alleviating its constraint. ✓ RPC's capacity to conserve methionine by providing methyl groups emerges as a conceivable mechanism, potentially contributing to the heightened milk protein yield. ✓ RPC supplementation led to an increase in DMI, consequently bolstering the availability of energy and nutrients at the onset of lactation. ✓ The observed surge in milk yield could also be attributed to the elevated energy and nutrient supply facilitated by increased DMI.	[190]
RPC	✓ Improved feed efficiency (FE) and average daily gain (ADG) and increased growth. ✓ Enhanced plasma glucose level. ✓ Elevated serum insulin-like growth factor-1 and suppressed expression of serum LPS-binding protein.	[191]
RPC	✓ Enhanced the biosynthesis of milk protein via the dephosphorylation of EEF2 and EIF2A during the negative nutrient balance in periparturient dairy cattle.	[192]
RPC	✓ Regulates the liver metabolism of periparturient dairy cows via activation of metabolic processes such as fatty acid synthesis and metabolism and glucose metabolism.	[193]

5. Conclusions

Altogether, addressing the challenges posed by nitrogen pollution while simultaneously enhancing milk production performance in dairy animals has become a crucial concern for sustainable livestock production. Another key concern is negative energy balance, which compromises the metabolism and health and consequent lactational performance of periparturient dairy cattle. This review article examined the significant roles of RPM and RPC in optimizing metabolism, nitrogen utilization, and improving milk production performance in dairy cattle. These strategies have emerged as powerful tools to mitigate the environmental impact of excessive nitrogen excretion while boosting the overall productivity of dairy animals. The incorporation of RPM into dairy cattle diet has been shown to have multifaceted benefits, ranging from its role as a limiting amino acid for milk production to its involvement in sulfur-containing amino acid synthesis and key metabolic pathways. Research has consistently demonstrated that their supplementation positively influences milk yield, milk protein synthesis, and overall metabolic efficiency. Similarly, RPC has been found to play a pivotal role in supporting metabolic processes during the transition period and beyond, contributing to improved liver health, lipid metabolism, and overall milk production performance.

Author Contributions: Conceptualization, B.H., M.Z.K., W.C. and C.W.; data collection, B.H., M.Z.K., W.C., Q.U., X.K., Y.C., H.L., A.K. and N.K.; writing—original draft preparation, B.H., M.Z.K. and C.W.; writing—review and editing, B.H., M.Z.K., W.C., Q.U., X.K., Y.C., H.L., A.K. and N.K.; visualization, B.H., M.Z.K., W.C., Q.U., X.K., Y.C., H.L., A.K. and N.K.; supervision, M.Z.K. and N.K. All authors have read and agreed to the published version of the manuscript.

Funding: This study was funded by the Shandong Province Modern Agricultural Technology System Donkey Industrial Innovation Team (grant number SDAIT-27), Livestock and Poultry Breeding Industry Project of the Ministry of Agriculture and Rural Affairs (grant number 19211162), Shandong Rural Revitalization Science and Technology Innovation Action Plan (Key Technology Innovation and Demonstration of Integrated Development of Dong-E Black Donkey Industry) (grant number

2021TZXD012), Open Project of Liaocheng University Animal Husbandry Discipline (grant number 319312101–14), Open Project of Shandong Collaborative Innovation Center for Donkey Industry Technology (grant number 3193308), Research on Donkey Pregnancy Improvement (grant number K20LC0901), and Liaocheng University Scientific Research Fund (grant number 318052025).

Acknowledgments: We express our sincere gratitude to the Liaocheng Research Institute of Donkey High-efficiency Breeding and Ecological Feeding, Liaocheng University for providing us with financial support.

Conflicts of Interest: The authors declare no conflict of interest.

References

1. Kabir, M.; Hasan, M.M.; Tanni, N.S.; Parvin, M.S.; Asaduzzaman, M.; Ehsan, M.A.; Islam, M.T. Metabolic profiling in periparturient dairy cows and its relation with metabolic diseases. *BMC Res. Notes* **2022**, *15*, 231. [CrossRef] [PubMed]
2. Khan, M.Z.; Liu, S.; Ma, Y.; Ma, M.; Ullah, Q.; Khan, I.M.; Wang, J.; Xiao, J.; Chen, T.; Khan, A.; et al. Overview of the effect of rumen-protected limiting amino acids (methionine and lysine) and choline on the immunity, antioxidative, and inflammatory status of periparturient ruminants. *Front. Immunol.* **2023**, *13*, 1042895. [CrossRef] [PubMed]
3. Janovick, N.A.; Drackley, J.K. Prepartum dietary management of energy intake affects postpartum intake and lactation performance by primiparous and multiparous Holstein cows. *J. Dairy Sci.* **2010**, *93*, 3086–3102. [CrossRef] [PubMed]
4. Janovick, N.A.; Boisclair, Y.R.; Drackley, J.K. Prepartum dietary energy intake affects metabolism and health during the periparturient period in primiparous and multiparous Holstein cows. *J. Dairy Sci.* **2011**, *94*, 1385–1400. [CrossRef] [PubMed]
5. Gumen, A.H.; Keskin, A.B.; Yilmazbas-Mecitoglu, G.U.; Karakaya, E.; Wiltbank, M.C. Dry period management and optimization of post-partum reproductive management in dairy cattle. *Reprod. Domest. Anim.* **2011**, *46*, 11–17. [CrossRef]
6. Xiao, J.; Khan, M.Z.; Ma, Y.; Alugongo, G.M.; Ma, J.; Chen, T.; Khan, A.; Cao, Z. The antioxidant properties of selenium and vitamin E; their role in periparturient dairy cattle health regulation. *Antioxidants* **2021**, *10*, 1555. [CrossRef]
7. Khan, M.Z.; Ma, Y.; Xiao, J.; Chen, T.; Ma, J.; Liu, S.; Wang, Y.; Khan, A.; Alugongo, G.M.; Cao, Z. Role of Selenium and Vitamins E and B9 in the Alleviation of Bovine Mastitis during the Periparturient Period. *Antioxidants* **2022**, *11*, 657. [CrossRef]
8. Khan, M.Z.; Zhang, Z.; Liu, L.; Wang, D.; Mi, S.; Liu, X.; Liu, G.; Guo, G.; Li, X.; Wang, Y.; et al. Folic acid supplementation regulates key immunity-associated genes and pathways during the periparturient period in dairy cows. *Asian-Australas. J. Anim. Sci.* **2020**, *33*, 1507. [CrossRef]
9. Sordillo, L.M.; Raphael, W. Significance of metabolic stress, lipid mobilization, and inflammation on transition cow disorders. *Vet. Clin. N. Am. Food Anim. Pract.* **2013**, *29*, 267–278. [CrossRef]
10. Drackley, J.K.; Dann, H.M.; Douglas, N.; Guretzky, N.A.; Litherland, N.B.; Underwood, J.P.; Loor, J.J. Physiological and pathological adaptations in dairy cows that may increase susceptibility to periparturient diseases and disorders. *Ital. J. Anim. Sci.* **2005**, *4*, 323–344. [CrossRef]
11. Li, C.; Batistel, F.; Osorio, J.S.; Drackley, J.K.; Luchini, D.; Loor, J.J. Peripartal rumen-protected methionine supplementation to higher energy diets elicits positive effects on blood neutrophil gene networks, performance, and liver lipid content in dairy cows. *J. Anim. Sci. Biotechnol.* **2016**, *7*, 18. [CrossRef]
12. Abreu, M.B.; Valldecabres, A.; Marcondes, M.I.; Correa, A.; Lobos, N.E.; Peterson, C.B.; Atwell, D.; Silva-del-Rio, N. Implications of supplementing mid-lactation multiparous Holstein cows fed high by-product low-forage diets with rumen-protected methionine and lysine in a commercial dairy. *Animal* **2023**, *17*, 100749. [CrossRef] [PubMed]
13. Batistel, F.; Arroyo, J.M.; Garces, C.I.; Trevisi, E.; Parys, C.; Ballou, M.A.; Cardoso, F.C.; Loor, J.J. Ethyl-cellulose rumen-protected methionine alleviates inflammation and oxidative stress and improves neutrophil function during the periparturient period and early lactation in Holstein dairy cows. *J. Dairy Sci.* **2018**, *101*, 480–490. [CrossRef] [PubMed]
14. Dai, H.; Coleman, D.N.; Hu, L.; Martinez-Cortés, I.; Wang, M.; Parys, C.; Shen, X.; Loor, J.J. Methionine and arginine supplementation alter inflammatory and oxidative stress responses during lipopolysaccharide challenge in bovine mammary epithelial cells in vitro. *J. Dairy Sci.* **2020**, *103*, 676–689. [CrossRef]
15. Arshad, U.; Husnain, A.; Poindexter, M.B.; Zimpel, R.; Nelson, C.D.; Santos, J.E. Rumen-protected choline reduces hepatic lipidosis by increasing hepatic triacylglycerol-rich lipoprotein secretion in dairy cows. *J. Dairy Sci.* **2023**, *in press*. [CrossRef]
16. Arshad, U.; Husnain, A.; Poindexter, M.B.; Zimpel, R.; Perdomo, M.C.; Santos, J.E. Effect of source and amount of rumen-protected choline on hepatic metabolism during induction of fatty liver in dairy cows. *J. Dairy Sci.* **2023**, *106*, 6860–6879. [CrossRef]
17. Toledo, M.Z.; Stangaferro, M.L.; Oliveira, R.C.; Monteiro, P.L., Jr.; Gennari, R.S.; Luchini, D.; Shaver, R.D.; Giordano, J.O.; Wiltbank, M.C. Effects of feeding rumen-protected methionine pre- and postpartum in multiparous Holstein cows: Health disorders and interactions with production and reproduction. *J. Dairy Sci.* **2023**, *106*, 2137–2152. [CrossRef]
18. Bell, M.J.; Wall, E.; Russell, G.; Simm, G.; Stott, A.W. The effect of improving cow productivity, fertility, and longevity on the global warming potential of dairy systems. *J. Dairy Sci.* **2011**, *94*, 3662–3678. [CrossRef]
19. Katongole, C.B.; Yan, T. Effect of varying dietary crude protein level on feed intake, nutrient digestibility, milk production, and nitrogen use efficiency by lactating Holstein-Friesian cows. *Animals* **2020**, *10*, 2439. [CrossRef]
20. Sinclair, K.D.; Garnsworthy, P.C.; Mann, G.E.; Sinclair, L.A. Reducing dietary protein in dairy cow diets: Implications for nitrogen utilization, milk production, welfare and fertility. *Animal* **2014**, *8*, 262–274. [CrossRef] [PubMed]

21. Wilkinson, J.M.; Garnsworthy, P.C. Impact of diet and fertility on greenhouse gas emissions and nitrogen efficiency of milk production. *Livestock* **2017**, *22*, 140–144. [CrossRef]
22. Prestegaard-Wilson, J.M.; Daley, V.L.; Drape, T.A.; Hanigan, M.D. A survey of United States dairy cattle nutritionists' practices and perceptions of reducing crude protein in lactating dairy cow diets. *Appl. Anim. Sci.* **2021**, *37*, 697–709. [CrossRef]
23. Place, S.E.; Mitloehner, F.M. The nexus of environmental quality and livestock welfare. *Annu. Rev. Anim. Biosci.* **2014**, *2*, 555–569. [CrossRef]
24. Abbasi, I.H.; Abbasi, F.; Abd El-Hack, M.E.; Abdel-Latif, M.A.; Soomro, R.N.; Hayat, K.; Mohamed, M.A.; Bodinga, B.M.; Yao, J.; Cao, Y. Critical analysis of excessive utilization of crude protein in ruminants ration: Impact on environmental ecosystem and opportunities of supplementation of limiting amino acids—A review. *Environ. Sci. Pollut. Res.* **2018**, *25*, 181–190. [CrossRef]
25. Aneja, V.P.; Schlesinger, W.H.; Erisman, J.W.; Behera, S.N.; Sharma, M.; Battye, W. Reactive nitrogen emissions from crop and livestock farming in India. *Atmos. Environ.* **2012**, *47*, 92–103. [CrossRef]
26. NRC. *Nutrient Requirements of Dairy Cattle*, 7th ed.; National Research Council National Academy Press: Washington, DC, USA, 2001.
27. Huhtanen, P.; Hristov, A.N. A meta-analysis of the effects of dietary protein concentration and degradability on milk protein yield and milk N efficiency in dairy cows. *J. Dairy Sci.* **2009**, *92*, 3222. [CrossRef]
28. Cabrita, A.R.; Dewhurst, R.J.; Melo, D.S.; Moorby, J.M.; Fonseca, A.J. Effects of dietary protein concentration and balance of absorbable amino acids on productive responses of dairy cows fed corn silage-based diets. *J. Dairy Sci.* **2011**, *94*, 4647–4656. [CrossRef] [PubMed]
29. Lee, C.; Hristov, A.N.; Heyler, K.S.; Cassidy, T.W.; Long, M.; Corl, B.A.; Karnati, S.K. Effects of dietary protein concentration and coconut oil supplementation on nitrogen utilization and production in dairy cows. *J. Dairy Sci.* **2011**, *94*, 5544–5557. [CrossRef] [PubMed]
30. Giallongo, F.; Hristov, A.N.; Oh, J.; Frederick, T.; Weeks, H.; Werner, J.; Lapierre, H.; Patton, R.A.; Gehman, A.; Parys, C. Effects of slow-release urea and rumen-protected methionine and histidine on performance of dairy cows. *J. Dairy Sci.* **2015**, *98*, 3292–3308. [CrossRef]
31. Wang, M.; Li, Y.; Yang, Z.; Shen, Y.; Cao, Y.; Li, Q.; Gao, Y.; Li, J. Effects of rumen-protected lysine and methionine supplementation in low-crude protein diets on lactation performance, nitrogen metabolism, rumen fermentation, and blood metabolites in Holstein cows. *Anim. Feed Sci. Technol.* **2022**, *292*, 115427. [CrossRef]
32. Lee, C.; Giallongo, F.; Hristov, A.N.; Lapierre, H.; Cassidy, T.W.; Heyler, K.S.; Varga, G.A.; Parys, C. Effect of dietary protein level and rumen-protected amino acid supplementation on amino acid utilization for milk protein in lactating dairy cows. *J. Dairy Sci.* **2015**, *98*, 1885–1902. [CrossRef]
33. Giallongo, F.; Harper, M.T.; Oh, J.; Lopes, J.C.; Lapierre, H.; Patton, R.A.; Parys, C.; Shinzato, I.; Hristov, A.N. Effects of rumen-protected methionine, lysine, and histidine on lactation performance of dairy cows. *J. Dairy Sci.* **2016**, *99*, 4437–4452. [CrossRef] [PubMed]
34. Schwab, C.G.; Broderick, G.A. A 100-Year Review: Protein and amino acid nutrition in dairy cows. *J. Dairy Sci.* **2017**, *100*, 10094–10112. [CrossRef] [PubMed]
35. Jayaprakash, G.; Sathiyabarathi, M.; Robert, M.A.; Tamilmani, T. Rumen-protected choline: A significance effect on dairy cattle nutrition. *Vet. World* **2016**, *9*, 837. [CrossRef]
36. Vailati-Riboni, M.; Zhou, Z.; Jacometo, C.B.; Minuti, A.; Trevisi, E.; Luchini, D.N.; Loor, J.J. Supplementation with rumen-protected methionine or choline during the transition period influences whole-blood immune response in periparturient dairy cows. *J. Dairy Sci.* **2017**, *100*, 3958–3968. [CrossRef] [PubMed]
37. Bollatti, J.M.; Zenobi, M.G.; Artusso, N.A.; Alfaro, G.F.; Lopez, A.M.; Barton, B.A.; Nelson, C.D.; Staples, C.R.; Santos, J.E. Timing of initiation and duration of feeding rumen-protected choline affects performance of lactating Holstein cows. *J. Dairy Sci.* **2020**, *103*, 4174–4191. [CrossRef] [PubMed]
38. Bollatti, J.M.; Zenobi, M.G.; Artusso, N.A.; Lopez, A.M.; Nelson, C.D.; Barton, B.A.; Staples, C.R.; Santos, J.E. Effects of rumen-protected choline on the inflammatory and metabolic status and health of dairy cows during the transition period. *J. Dairy Sci.* **2020**, *103*, 4192–4205. [CrossRef]
39. Pereira, A.B.; Moura, D.C.; Whitehouse, N.L.; Brito, A.F. Production and nitrogen metabolism in lactating dairy cows fed finely ground field pea plus soybean meal or canola meal with or without rumen-protected methionine supplementation. *J. Dairy Sci.* **2020**, *103*, 3161–3176. [CrossRef]
40. Broderick, G.A.; Faciola, A.P.; Armentano, L.E. Replacing dietary soybean meal with canola meal improves production and efficiency of lactating dairy cows. *J. Dairy Sci.* **2015**, *98*, 5672–5687. [CrossRef]
41. Pan, F.; Li, P.; Hao, G.; Liu, Y.; Wang, T.; Liu, B. Enhancing Milk Production by Nutrient Supplements: Strategies and Regulatory Pathways. *Animals* **2023**, *13*, 419. [CrossRef]
42. Qi, H.; Wang, L.; Zhang, M.; Wang, Z.; Gao, X.; Li, M. Methionine and leucine induce ARID1A degradation to promote mTOR expression and milk synthesis in mammary epithelial cells. *J. Nutr. Biochem.* **2022**, *101*, 108924. [CrossRef]
43. Wang, F.; Shi, H.; Wang, S.; Wang, Y.; Cao, Z.; Li, S. Amino acid metabolism in dairy cows and their regulation in milk synthesis. *Curr. Drug Metab.* **2019**, *20*, 36–45. [CrossRef] [PubMed]
44. Han, M.; Zhang, M. The regulatory mechanism of amino acids on milk protein and fat synthesis in mammary epithelial cells: A mini review. *Anim. Biotechnol.* **2023**, *34*, 402–412. [CrossRef] [PubMed]

45. Li, P.; Zhou, C.; Li, X.; Yu, M.; Li, M.; Gao, X. CRTC2 is a key mediator of amino acid-induced milk fat synthesis in mammary epithelial cells. *J. Agric. Food Chem.* **2019**, *67*, 10513–10520. [CrossRef]
46. Huo, N.; Yu, M.; Li, X.; Zhou, C.; Jin, X.; Gao, X. PURB is a positive regulator of amino acid-induced milk synthesis in bovine mammary epithelial cells. *J. Cell Physiol.* **2019**, *234*, 6992–7003. [CrossRef] [PubMed]
47. Nan, X.; Bu, D.; Li, X.; Wang, J.; Wei, H.; Hu, H.; Zhou, L.; Loor, J.J. Ratio of lysine to methionine alters expression of genes involved in milk protein transcription and translation and mTOR phosphorylation in bovine mammary cells. *Physiol. Genom.* **2014**, *46*, 268–275. [CrossRef] [PubMed]
48. Van Amburgh, M.E.; VanCollao-Saenz, E.A.; Higgs, R.J.; Ross, D.A.; Recktenwald, E.B.; Raffrenato, E.; Chase, L.E.; Overton, T.R.; Mills, J.K.; Foskolos, A. The Cornell Net Carbohydrate and Protein System: Updates to the model and evaluation of version 6.5. *J. Dairy Sci.* **2015**, *98*, 6361–6380. [CrossRef]
49. Elsaadawy, S.A.; Wu, Z.; Wang, H.; Hanigan, M.D.; Bu, D. Supplementing ruminally protected lysine, methionine, or combination improved milk production in transition dairy cows. *Front. Vet. Sci.* **2022**, *9*, 780637. [CrossRef]
50. Elsaadawy, S.A.; Wu, Z.; Bu, D. Feasibility of Supplying Ruminally Protected Lysine and Methionine to Periparturient Dairy Cows on the Efficiency of Subsequent Lactation. *Front. Vet. Sci.* **2022**, *9*, 892709. [CrossRef]
51. Chen, Z.H.; Broderick, G.A.; Luchini, N.D.; Sloan, B.K.; Devillard, E. Effect of feeding different sources of rumen-protected methionine on milk production and N-utilization in lactating dairy cows. *J. Dairy Sci.* **2011**, *94*, 1978–1988. [CrossRef]
52. Osorio, J.S.; Ji, P.; Drackley, J.K.; Luchini, D.; Loor, J.J. Supplemental Smartamine M or MetaSmart during the transition period benefits postpartal cow performance and blood neutrophil function. *J. Dairy Sci.* **2013**, *96*, 6248–6263. [CrossRef] [PubMed]
53. Brosnan, J.T.; Brosnan, M.E. The sulfur-containing amino acids: An overview. *J. Nutr.* **2006**, *136*, 1636S–1640S. [CrossRef]
54. Bin, P.; Huang, R.; Zhou, X. Oxidation resistance of the sulfur amino acids: Methionine and Cysteine. *Biomed Res. Int.* **2017**, *2017*, 9584932. [CrossRef] [PubMed]
55. Humer, E.; Bruggeman, G.; Zebeli, Q. A meta-analysis on the impact of the supplementation of rumen-protected choline on the metabolic health and performance of dairy cattle. *Animals* **2019**, *9*, 566. [CrossRef] [PubMed]
56. Arshad, U.; Zenobi, M.G.; Staples, C.R.; Santos, J.E. Meta-analysis of the effects of supplemental rumen-protected choline during the transition period on performance and health of parous dairy cows. *J. Dairy Sci.* **2020**, *103*, 282–300. [CrossRef] [PubMed]
57. Lopreiato, V.; Alharthi, A.S.; Liang, Y.; Elolimy, A.A.; Bucktrout, R.; Socha, M.T.; Trevisi, E.; Loor, J.J. Influence of Cobalt Source, Folic Acid, and Rumen-Protected Methionine on Performance, Metabolism, and Liver Tissue One-Carbon Metabolism Biomarkers in Peripartal Holstein Cows. *Animals* **2023**, *13*, 2107. [CrossRef]
58. Aboragah, A.A.; Alharthi, A.S.; Wichasit, N.; Loor, J.J. Body condition prepartum and its association with term placentome nutrient transporters, one carbon metabolism pathway activity, and intermediate metabolites in Holstein cows. *Res. Vet. Sci.* **2023**, *162*, 104956. [CrossRef] [PubMed]
59. Jiang, Q.; Sherlock, D.N.; Zhang, H.; Guyader, J.; Pan, Y.X.; Loor, J.J. One-carbon metabolism and related pathways in ruminal and small intestinal epithelium of lactating dairy cows. *J. Anim. Sci.* **2023**, *101*, skad062. [CrossRef]
60. Palombo, V.; Alharthi, A.; Batistel, F.; Parys, C.; Guyader, J.; Trevisi, E.; D'Andrea, M.; Loor, J.J. Unique Adaptations in Neonatal Hepatic Transcriptome, Nutrient Signaling, and One-Carbon Metabolism in Response to Feeding Ethyl Cellulose Rumen-Protected Methionine during Late-Gestation in Holstein Cows. *BMC Genom.* **2021**, *22*, 280. [CrossRef]
61. McFadden, J.W.; Girard, C.L.; Tao, S.; Zhou, Z.; Bernard, J.K.; Duplessis, M.; White, H.M. Symposium Review: One-Carbon Metabolism and Methyl Donor Nutrition in the Dairy Cow. *J. Dairy Sci.* **2020**, *103*, 5668–5683. [CrossRef]
62. Zhou, Z.; Vailati-Riboni, M.; Luchini, D.N.; Loor, J.J. Methionine and Choline Supply during the Periparturient Period Alter Plasma Amino Acid and One-Carbon Metabolism Profiles to Various Extents: Potential Role in Hepatic Metabolism and Antioxidant Status. *Nutrients* **2016**, *9*, 10. [CrossRef]
63. Clare, C.E.; Brassington, A.H.; Kwong, W.Y.; Sinclair, K.D. One-Carbon Metabolism: Linking Nutritional Biochemistry to Epigenetic Programming of Long-Term Development. *Annu. Rev. Anim. Biosci.* **2019**, *7*, 263–287. [CrossRef] [PubMed]
64. Zhang, N. Role of Methionine on Epigenetic Modification of DNA Methylation and Gene Expression in Animals. *Anim. Nutr.* **2018**, *4*, 11–16. [CrossRef] [PubMed]
65. Obeid, R.; Herrmann, W. Hcy and Lipids: S-Adenosyl Methionine as a Key Intermediate. *FEBS Lett.* **2009**, *583*, 1215–1225. [CrossRef] [PubMed]
66. Zhou, Z.; Trevisi, E.; Luchini, D.N.; Loor, J.J. Differences in liver functionality indexes in peripartal dairy cows fed rumen-protected methionine or choline are associated with performance, oxidative stress status, and plasma amino acid profiles. *J. Dairy Sci.* **2017**, *100*, 6720–6732. [CrossRef] [PubMed]
67. Zhou, Y.F.; Zhou, Z.; Batistel, F.; Martinez-Cortés, I.; Pate, R.T.; Luchini, D.L.; Loor, J.J. Methionine and choline supply alter transmethylation, transsulfuration, and cytidine 5′-diphosphocholine pathways to different extents in isolated primary liver cells from dairy cows. *J. Dairy Sci.* **2018**, *101*, 11384–11395. [CrossRef]
68. Ardalan, M.; Rezayazdi, K.; Dehghan-Banadaky, M. Effect of rumen-protected choline and methionine on physiological and metabolic disorders and reproductive indices of dairy cows. *J. Anim. Physiol. Anim. Nutr.* **2010**, *94*, e259–e265. [CrossRef] [PubMed]
69. Obeid, R. The metabolic burden of methyl donor deficiency with focus on the betaine Hcy methyltransferase pathway. *Nutrients* **2013**, *5*, 3481–3495. [CrossRef]

70. Potts, S.B.; Brady, K.M.; Scholte, C.M.; Moyes, K.M.; Sunny, N.E.; Erdman, R.A. Rumen-protected choline and methionine during the periparturient period affect choline metabolites, amino acids, and hepatic expression of genes associated with one-carbon and lipid metabolism. *J. Dairy Sci.* **2023**, *7*, 4559–4579. [CrossRef]
71. Hirabayashi, T.; Kawaguchi, M.; Harada, S.; Mouri, M.; Takamiya, R.; Miki, Y.; Sato, H.; Taketomi, Y.; Yokoyama, K.; Kobayashi, T.; et al. Hepatic phosphatidylcholine catabolism driven by PNPLA7 and PNPLA8 supplies endogenous choline to replenish the methionine cycle with methyl groups. *Cell Rep.* **2023**, *42*, 2. [CrossRef]
72. Martínez, Y.; Li, X.; Liu, G.; Bin, P.; Yan, W.; Más, D.; Valdivié, M.; Hu, C.A.; Ren, W.; Yin, Y. The role of methionine on metabolism, oxidative stress, and diseases. *Amino Acids* **2017**, *49*, 2091–2098. [CrossRef] [PubMed]
73. Crouse, M.S.; Freetly, H.C.; Lindholm-Perry, A.K.; Neville, B.W.; Oliver, W.T.; Lee, R.T.; Syring, J.G.; King, L.E.; Reynolds, L.P.; Dahlen, C.R.; et al. One-carbon metabolite supplementation to heifers for the first 14 d of the estrous cycle alters the plasma and hepatic one-carbon metabolite pool and methionine-folate cycle enzyme transcript abundance in a dose-dependent manner. *J. Anim. Sci.* **2023**, *13*, skac419. [CrossRef]
74. Liu, Y.Q.; Wang, C.; Liu, C.; Zhang, J.; Liu, Q. Effects of coated folic acid and coated methionine on growth performance, nutrient digestibility and rumen fermentation in Simmental bulls. *Anim. Feed Sci. Technol.* **2023**, *298*, 115596. [CrossRef]
75. Seymour, W.M. Role of methionine and methionine precursors in transition cow nutrition with emphasis on liver function. In Proceedings of the 2016 Florida Ruminant Nutrition Symposium, Gainesville, FL, USA, 15–17 February 2016; pp. 11–16.
76. Mackay, D.S.; Brophy, J.D.; McBreairty, L.E.; McGowan, R.A.; Bertolo, R.F. Intrauterine growth restriction leads to changes in sulfur amino acid metabolism, but not global DNA methylation. Yucatan miniature piglets. *J. Nutr. Biochem.* **2012**, *23*, 1121–1127. [CrossRef] [PubMed]
77. Troen, A.M.; Lutgens, E.; Smith, D.E.; Rosenberg, I.H.; Selhub, J. The atherogenic effect of excess methionine intake. *Proc. Natl. Acad. Sci. USA* **2003**, *100*, 15089–15094. [CrossRef]
78. Rezzi, S.; Ramadan, Z.; Fay, L.B.; Kochhar, S. Nutritional metabonomics: Applications and perspectives. *J. Proteome Res.* **2007**, *6*, 513–525. [CrossRef]
79. Nicholson, J.K.; Lindon, J.C.; Holmes, E. 'Metabonomics': Understanding the metabolic responses of living systems to pathophysiological stimuli via multivariate statistical analysis of biological NMR spectroscopic data. *Xenobiotica* **2008**, *29*, 1181–1189. [CrossRef]
80. Bauchart-Thevret, C.; Stoll, B.; Chacko, S.; Burrin, D.G. Sulfur amino acid deficiency upregulates intestinal methionine cycle activity and suppresses epithelial growth in neonatal pigs. *Am. J. Physiol.-Endocrinol. Metab.* **2009**, *296*, E1239–E1250. [CrossRef]
81. Riedijk, M.A.; Stoll, B.; Chacko, S.; Schierbeek, H.; Sunehag, A.L.; van Goudoever, J.B.; Burrin, D.G. Methionine transmethylation and transsulfuration in the piglet gastrointestinal tract. *Proc. Natl. Acad. Sci. USA* **2007**, *104*, 3408–3413. [CrossRef]
82. Shoveller, A.K.; Brunton, J.A.; Pencharz, P.B.; Ball, R.O. The methionine requirement is lower in neonatal piglets fed parenterally than in those fed enterally. *J. Nutr.* **2003**, *133*, 1390–1397. [CrossRef]
83. Blachier, F.; Mariotti, F.; Huneau, J.F.; Tomé, D. Effects of amino acid-derived luminal metabolites on the colonic epithelium and physiopathological consequences. *Amino Acids* **2007**, *33*, 547–562. [CrossRef] [PubMed]
84. Swain, B.K.; Johri, T.S. Effect of supplemental methionine, choline and their combinations on the performance and immune response of broilers. *Br. Poult. Sci.* **2000**, *41*, 83–88. [CrossRef] [PubMed]
85. Carew, L.B.; McMurtry, J.P.; Alster, F.A. Effect of methionine deficiencies on plasma levels of thyroid hormones insulin-like growth factors-I and -II, liver and body weights, and feed intake in growing chickens. *Poult. Sci.* **2003**, *82*, 1932–1938. [CrossRef] [PubMed]
86. D'Mello, J.P. Responses of growing poultry to amino acids. *Amino Acids Anim. Nutr.* **2003**, 237–263. [CrossRef]
87. Lee, C.; Hristov, A.N.; Cassidy, T.W.; Heyler, K.S.; Lapierre, H.; Varga, G.A.; De Veth, M.J.; Patton, R.A.; Parys, C. Rumen-protected lysine, methionine, and histidine increase milk protein yield in dairy cows fed a metabolizable protein-deficient diet. *J. Dairy Sci.* **2012**, *95*, 6042–6056. [CrossRef]
88. Van den Bossche, T.; Goossens, K.; Ampe, B.; Haesaert, G.; De Sutter, J.; De Boever, J.L.; Vandaele, L. Effect of supplementing rumen-protected methionine, lysine, and histidine to low-protein diets on the performance and nitrogen balance of dairy cows. *J. Dairy Sci.* **2023**, *13*, 1790–1802. [CrossRef]
89. Wang, W.; Ye, L.; Dou, X.; Liu, H.; Han, D. Effects of Rumen-Protected Methionine Supplementation on Growth Performance, Nutrient Digestion, Nitrogen Utilisation and Plasma Amino Acid Profiles of Liaoning Cashmere Goats. *Animals* **2023**, *13*, 2995. [CrossRef]
90. Zou, S.; Ji, S.; Xu, H.; Wang, M.; Li, B.; Shen, Y.; Li, Y.; Gao, Y.; Li, J.; Cao, Y.; et al. Rumen-Protected Lysine and Methionine Supplementation Reduced Protein Requirement of Holstein Bulls by Altering Nitrogen Metabolism in Liver. *Animals* **2023**, *13*, 843. [CrossRef]
91. Li, Y.; Wei, J.; Dou, M.; Liu, S.; Yan, B.; Li, C.; Khan, M.Z.; Zhang, Y.; Xiao, J. Effects of rumen-protected methionine supplementation on production performance, apparent digestibility, blood parameters, and ruminal fermentation of lactating Holstein dairy cows. *Front. Vet. Sci.* **2022**, *9*, 981757. [CrossRef]
92. Pereira, A.B.; Whitehouse, N.L.; Aragona, K.M.; Schwab, C.S.; Reis, S.F.; Brito, A.F. Production and nitrogen utilization in lactating dairy cows fed ground field peas with or without ruminally protected lysine and methionine. *J. Dairy Sci.* **2017**, *100*, 6239–6255. [CrossRef]

93. Ding, L.; Shen, Y.; Wang, Y.; Zhou, G.; Zhang, X.; Wang, M.; Loor, J.J.; Chen, L.; Zhang, J. Jugular arginine supplementation increases lactation performance and nitrogen utilization efficiency in lactating dairy cows. *J. Anim. Sci. Biotechnol.* **2019**, *10*, 3. [CrossRef] [PubMed]
94. Noftsger, S.; St-Pierre, N.R. Supplementation of methionine and selection of highly digestible rumen undegradable protein to improve nitrogen efficiency for milk production. *J. Dairy Sci.* **2003**, *86*, 958–969. [CrossRef] [PubMed]
95. Wang, C.; Liu, H.Y.; Wang, Y.M.; Yang, Z.Q.; Liu, J.X.; Wu, Y.M.; Yan, T.; Ye, H.W. Effects of dietary supplementation of methionine and lysine on milk production and nitrogen utilization in dairy cows. *J. Dairy Sci.* **2010**, *93*, 3661–3670. [CrossRef]
96. Sobhanirad, S.; Carlson, D.; Bahari Kashani, R. Effect of zinc methionine or zinc sulfate supplementation on milk production and composition of milk in lactating dairy cows. *Biol. Trace Elem. Res.* **2010**, *136*, 48–54. [CrossRef]
97. Kellogg, D.W.; Tomlinson, D.J.; Socha, M.T.; Johnson, A.B. Effects of zinc methionine complex on milk production and somatic cell count of dairy cows: Twelve-trial summary. *Prof. Anim. Sci.* **2004**, *20*, 295–301. [CrossRef]
98. Drackley, J.K. Biology of dairy cows during the transition period: The final frontier? *J. Dairy Sci.* **1999**, *82*, 2259–2273. [CrossRef] [PubMed]
99. Ma, Y.F.; Batistel, F.; Xu, T.L.; Han, L.Q.; Bucktrout, R.; Liang, Y.; Coleman, D.N.; Parys, C.; Loor, J.J. Phosphorylation of AKT serine/threonine kinase and abundance of milk protein synthesis gene networks in mammary tissue in response to the supply of methionine in periparturient Holstein cows. *J. Dairy Sci.* **2019**, *102*, 4264–4274. [CrossRef] [PubMed]
100. Lean, I.J.; Ondarza, M.B.; Sniffen, C.J.; Santos, J.P.; Griswold, K.E. Meta-analysis to predict the effects of metabolizable amino acids on dairy cattle performance. *J. Dairy Sci.* **2018**, *101*, 340–364. [CrossRef]
101. Batistel, F.; Arroyo, J.M.; Bellingeri, A.; Wang, L.; Saremi, B.; Parys, C.; Trevisi, E.; Cardoso, F.C.; Loor, J.J. Ethyl-cellulose rumen-protected methionine enhances performance during the periparturient period and early lactation in Holstein dairy cows. *J. Dairy Sci.* **2017**, *100*, 7455–7467. [CrossRef]
102. Bionaz, M.; Loor, J.J. Gene networks driving bovine mammary protein synthesis during the lactation cycle. *Bioinform. Biol. Insights* **2011**, *5*, BBI-S7003. [CrossRef]
103. Loor, J.J.; Bionaz, M.; Drackley, J.K. Systems physiology in dairy cattle: Nutritional genomics and beyond. *Annu. Rev. Anim. Biosci.* **2013**, *1*, 365–392. [CrossRef] [PubMed]
104. Wang, J.H.; Diao, Q.Y.; Xu, X.C.; Tu, Y.; Zhang, N.F.; Yun, Q. The limiting sequence and proper ratio of lysine, methionine, and threonine for calves fed milk replacers containing soy protein. *Asian-Australas. J. Anim. Sci.* **2012**, *25*, 224–233. [CrossRef]
105. Li, Y.; Bi, Y.; Diao, Q.; Piao, M.; Wang, B.; Kong, F.; Hu, F.; Tang, M.; Sun, Y.; Tu, Y. The limiting sequence and appropriate amino acid ratio of lysine, methionine, and threonine for seven-to-nine-month-old Holstein heifers fed corn–soybean meal-based diet. *Animals* **2019**, *9*, 750. [CrossRef]
106. Lee, C.; Hristov, A.N.; Heyler, K.S.; Cassidy, T.W.; Lapierre, H.; Varga, G.A.; Parys, C. Effects of metabolizable protein supply and amino acid supplementation on nitrogen utilization, milk production, and ammonia emissions from manure in dairy cows. *J. Dairy Sci.* **2012**, *95*, 5253–5268. [CrossRef]
107. Park, J.K.; Yeo, J.M.; Bae, G.S.; Kim, E.J.; Kim, C.H. Effects of supplementing limiting amino acids on milk production in dairy cows consuming a corn grain and soybean meal-based diet. *J. Anim. Sci. Technol.* **2020**, *62*, 485. [CrossRef]
108. Carder, E.G.; Weiss, W.P. Short- and longer-term effects of feeding increased metabolizable protein with or without an altered amino acid profile to dairy cows immediately postpartum. *J. Dairy Sci.* **2017**, *100*, 4528–4538. [CrossRef] [PubMed]
109. Socha, M.T.; Putnam, D.E.; Garthwaite, B.D.; Whitehouse, N.L.; Kierstead, N.A.; Schwab, C.G.; Ducharme, G.A.; Robert, J.C. Improving intestinal amino acid supply of pre- and postpartum dairy cows with rumen-protected methionine and lysine. *J. Dairy Sci.* **2005**, *88*, 1113–1126. [CrossRef] [PubMed]
110. Xu, S.; Harrison, J.H.; Chalupa, W.; Sniffen, C.; Julien, W.; Sato, H.; Fujieda, T.; Watanabe, K.; Ueda, T.; Suzuki, H. The effect of ruminal bypass lysine and methionine on milk yield and composition of lactating cows. *J. Dairy Sci.* **1998**, *81*, 1062–1077. [CrossRef]
111. Zhou, Z.; Vailati-Riboni, M.; Trevisi, E.; Drackley, J.K.; Luchini, D.N.; Loor, J.J. Better postpartal performance in dairy cows supplemented with rumen-protected methionine compared with choline during the peripartal period. *J. Dairy Sci.* **2016**, *99*, 8716–8732. [CrossRef]
112. Lee, C.; Lobos, N.E.; Weiss, W.P. Effects of supplementing rumen-protected lysine and methionine during prepartum and postpartum periods on performance of dairy cows. *J. Dairy Sci.* **2019**, *102*, 11026–11039. [CrossRef]
113. Weiss, W.P. Effects of feeding diets composed of corn silage and a corn milling product with and without supplemental lysine and methionine to dairy cows. *J. Dairy Sci.* **2019**, *102*, 2075–2084. [CrossRef]
114. Yu, Y.; Zhen, Z.; Qi, H.; Yuan, X.; Gao, X.; Zhang, M. U2AF65 enhances milk synthesis and growth of bovine mammary epithelial cells by positively regulating the mTOR-SREBP-1c signaling pathway. *Cell Biochem. Funct.* **2019**, *37*, 93–101. [CrossRef]
115. Zhang, M.; Chen, D.; Zhen, Z.; Ao, J.; Yuan, X.; Gao, X. Annexin A2 positively regulates milk synthesis and proliferation of bovine mammary epithelial cells through the mTOR signaling pathway. *J. Cell. Physiol.* **2018**, *233*, 2464–2475. [CrossRef]
116. Liu, Y.; Wang, X.; Zhen, Z.; Yu, Y.; Qiu, Y.; Xiang, W. GRP78 regulates milk biosynthesis and the proliferation of bovine mammary epithelial cells through the mTOR signaling pathway. *Cell. Mol. Biol. Lett.* **2019**, *24*, 2. [CrossRef]
117. Huang, X.; Zang, Y.; Zhang, M.; Yuan, X.; Li, M.; Gao, X. Nuclear factor of κB1 is a key regulator for the transcriptional activation of milk synthesis in bovine mammary epithelial cells. *DNA Cell Biol.* **2017**, *36*, 295–302. [CrossRef] [PubMed]

118. Zhen, Z.; Zhang, M.; Yuan, X.; Qu, B.; Yu, Y.; Gao, X.; Qiu, Y. DEAD-box helicase 6 (DDX6) is a new negative regulator for milk synthesis and proliferation of bovine mammary epithelial cells. *In Vitro Cell. Dev. Biol. Anim.* **2018**, *54*, 52–60. [CrossRef]
119. Li, P.; Yu, M.; Zhou, C.; Qi, H.; Wen, X.; Hou, X.; Li, M.; Gao, X. FABP5 is a critical regulator of methionine-and estrogen-induced SREBP-1c gene expression in bovine mammary epithelial cells. *J. Cell. Physiol.* **2019**, *234*, 537–549. [CrossRef]
120. Zhou, Y.; Zhou, Z.; Peng, J.; Loor, J.J. Methionine and valine activate the mammalian target of rapamycin complex 1 pathway through heterodimeric amino acid taste receptor (TAS1R1/TAS1R3) and intracellular Ca^{2+} in bovine mammary epithelial cells. *J. Dairy Sci.* **2018**, *101*, 11354–11363. [CrossRef] [PubMed]
121. Zhou, Z.; Bulgari, O.; Vailati-Riboni, M.; Trevisi, E.; Ballou, M.A.; Cardoso, F.C.; Luchini, D.N.; Loor, J.J. Rumen-protected methionine compared with rumen-protected choline improves immunometabolic status in dairy cows during the peripartal period. *J. Dairy Sci.* **2016**, *99*, 8956–8969. [CrossRef]
122. Zanton, G.I.; Bowman, G.R.; Vázquez-Añón, M.; Rode, L.M. Meta-analysis of lactation performance in dairy cows receiving supplemental dietary methionine sources or postruminal infusion of methionine. *J. Dairy Sci.* **2014**, *97*, 7085–7101. [CrossRef] [PubMed]
123. Junior, V.C.; Lopes, F.; Schwab, C.G.; Toledo, M.Z.; Collao-Saenz, E.A. Effects of rumen-protected methionine supplementation on the performance of high production dairy cows in the tropics. *PLoS ONE* **2021**, *16*, e0243953. [CrossRef]
124. Danesh Mesgaran, M.; Kargar, H.; Janssen, R.; Danesh Mesgaran, S.; Ghesmati, A.; Vatankhah, A. Rumen-protected zinc–methionine dietary inclusion alters dairy cow performances, and oxidative and inflammatory status under long-term environmental heat stress. *Front. Vet. Sci.* **2022**, *9*, 935939. [CrossRef]
125. Sun, F.; Cao, Y.; Cai, C.; Li, S.; Yu, C.; Yao, J. Regulation of nutritional metabolism in transition dairy cows: Energy homeostasis and health in response to post-ruminal choline and methionine. *PLoS ONE* **2016**, *11*, e0156770. [CrossRef]
126. Sun, P.; Wang, J.; Liu, W.; Bu, D.P.; Liu, S.J.; Zhang, K.Z. Hydroxy-selenomethionine: A novel organic selenium source that improves antioxidant status and selenium concentrations in milk and plasma of mid-lactation dairy cows. *J. Dairy Sci.* **2017**, *100*, 9602–9610. [CrossRef] [PubMed]
127. Li, Y.; Liu, J.X.; Xiong, J.L.; Wang, Y.M.; Zhang, W.X.; Wang, D.M. Effect of hydroxyselenomethionine on lactation performance, blood profiles, and transfer efficiency in early-lactating dairy cows. *J. Dairy Sci.* **2019**, *102*, 6167–6173. [CrossRef] [PubMed]
128. Liang, S.L.; Wei, Z.H.; Wu, J.J.; Dong, X.L.; Liu, J.X.; Wang, D.M. Effect of N-acetyl-l-methionine supplementation on lactation performance and plasma variables in mid-lactating dairy cows. *J. Dairy Sci.* **2019**, *102*, 5182–5190. [CrossRef]
129. Amaro, F.X.; Kim, D.; Restelatto, R.; Carvalho, P.; Arriola, K.; Duvalsaint, E.J.; Cervantes, A.P.; Jiang, Y.; Agarussi, M.C.; Silva, V.P.; et al. Lactational performance of dairy cows in response to supplementing N-acetyl-l-methionine as a source of rumen-protected methionine. *J. Dairy Sci.* **2022**, *105*, 2301–2314. [CrossRef]
130. Fagundes, M.A.; Yang, S.Y.; Eun, J.S.; Hall, J.O.; Moon, J.O.; Park, J.S. Influence of supplementing a methionine derivative, N-acetyl-l-methionine, in dairy diets on production and ruminal fermentation by lactating cows during early to mid lactation. *J. Dairy Sci.* **2018**, *101*, 7082–7094. [CrossRef]
131. Räisänen, S.E.; Zhu, X.; Zhou, C.; Lage, C.F.; Fetter, M.; Silvestre, T.; Stefenoni, H.; Wasson, D.E.; Cueva, S.F.; Eun, J.S.; et al. Production effects and bioavailability of N-acetyl-l-methionine in lactating dairy cows. *J. Dairy Sci.* **2022**, *105*, 313–328. [CrossRef] [PubMed]
132. Alharthi, A.S.; Coleman, D.N.; Liang, Y.; Batistel, F.; Elolimy, A.A.; Yambao, R.C.; Abdel-Hamied, E.; Pan, Y.X.; Parys, C.; Alhidary, I.A.; et al. Hepatic 1-carbon metabolism enzyme activity, intermediate metabolites, and growth in neonatal Holstein dairy calves are altered by maternal supply of methionine during late pregnancy. *J. Dairy Sci.* **2019**, *102*, 10291–10303. [CrossRef] [PubMed]
133. Noftsger, S.; St-Pierre, N.R.; Sylvester, J.T. Determination of rumen degradability and ruminal effects of three sources of methionine in lactating cows. *J. Dairy Sci.* **2005**, *88*, 223–237. [CrossRef] [PubMed]
134. Ardalan, M.; Dehghan-Banadaky, M.; Rezayazdi, K.; Hossein-Zadeh, N.G. The effect of rumen-protected methionine and choline on plasma metabolites of Holstein dairy cows. *J. Agric. Sci.* **2011**, *149*, 639–646. [CrossRef]
135. Abedal-Majed, M.A.; Titi, H.H.; Al-Qaisi, M.; Abdelqader, A.; Tabbaa, M.J. The effects of rumen protected methionine supplementation on the performance of primiparous dairy cows using the Presynch-Ovsynch protocol. *Anim. Sci. J.* **2023**, *94*, e13835. [CrossRef]
136. Irawan, A.; Sofyan, A.; Wahyono, T.; Harahap, M.A.; Febrisiantosa, A.; Herdian, H.; Sakti, A.A.; Jayanegara, A. Relationships between dietary rumen-protected lysine and methionine with the lactational performance of dairy cows: A meta-analysis. *Anim. Biosci.* **2023**, *13*, 1–19. [CrossRef] [PubMed]
137. Melendez, P.; Möller, J.; Arevalo, A.; Pinedo, P. The effect of rumen-protected lysine and methionine on milk yield, milk components, and body weight in grazing Holstein cows during spring calving season in the southern hemisphere. *Livest. Sci.* **2023**, *272*, 105230. [CrossRef]
138. Pitkänen, O.; Halmemies-Beauchet-Filleau, A.; Räisänen, S.E.; Jaakkola, S.; Kokkonen, T.; Vanhatalo, A. Processed fava bean as a substitute for rapeseed meal with or without rumen-protected methionine supplement in grass silage-based dairy cow diets. *J. Dairy Sci.* **2023**, *106*, 3217–3232. [CrossRef]
139. Richards, A.T.; Knapp, J.R.; Summer, P.; Ohta, Y.; Boerman, J.P. Bioavailability of a novel rumen protected methionine supplement and its effect on milk production and body composition in dairy cows. *Anim. Feed Sci. Technol.* **2023**, *304*, 115750. [CrossRef]
140. Wei, C.; He, T.; Wan, X.; Liu, S.; Dong, Y.; Qu, Y. Meta-Analysis of Rumen-Protected Methionine in Milk Production and Composition of Dairy Cows. *Animals* **2022**, *12*, 1505. [CrossRef]

141. Li, L.; Liu, L.; Qu, B.; Li, X.; Gao, X.; Zhang, M. Twinfilin 1 enhances milk biosynthesis and proliferation of bovine mammary epithelial cells via the mTOR signaling pathway. *Biochem. Biophys. Res. Commun.* **2017**, *492*, 289–294. [CrossRef]
142. Wang, C.; Zhao, F.; Liu, J.; Liu, H. Dipeptide (methionyl-methionine) transport and its effect on β-casein synthesis in bovine mammary epithelial cells. *Cell. Physiol. Biochem.* **2018**, *49*, 479–488. [CrossRef] [PubMed]
143. Yang, J.X.; Wang, C.H.; Xu, Q.B.; Zhao, F.Q.; Liu, J.X.; Liu, H.Y. Methionyl-Methionine promotes α-s1 casein synthesis in bovine mammary gland explants by enhancing intracellular substrate availability and activating JAK2-STAT5 and mTOR-mediated signaling pathways. *J. Nutr.* **2015**, *145*, 1748–1753. [CrossRef] [PubMed]
144. Gao, H.N.; Zhao, S.G.; Zheng, N.; Zhang, Y.D.; Wang, S.S.; Zhou, X.Q.; Wang, J.Q. Combination of histidine, lysine, methionine, and leucine promotes β-casein synthesis via the mechanistic target of rapamycin signaling pathway in bovine mammary epithelial cells. *J. Dairy Sci.* **2017**, *100*, 7696–7709. [CrossRef] [PubMed]
145. Duan, X.; Liu, C.; Gong, X.; Yang, Y.; Jiao, H.; Lin, Y.; Hou, X. mTORC2-AKT-LAT1 Signalling Participates in Methionine-induced β-CASEIN Expression in Mammary Epithelial Cells of Dairy Cows. *J. Anim. Physiol. Anim. Nutr.* **2023**, *in press*. [CrossRef]
146. Dong, X.; Zhou, Z.; Saremi, B.; Helmbrecht, A.; Wang, Z.; Loor, J.J. Varying the ratio of Lys: Met while maintaining the ratios of Thr: Phe, Lys: Thr, Lys: His, and Lys: Val alters mammary cellular metabolites, mammalian target of rapamycin signaling, and gene transcription. *J. Dairy Sci.* **2018**, *101*, 1708–1718. [CrossRef]
147. Apelo, S.A.; Singer, L.M.; Lin, X.Y.; McGilliard, M.L.; St-Pierre, N.R.; Hanigan, M.D. Isoleucine, leucine, methionine, and threonine effects on mammalian target of rapamycin signaling in mammary tissue. *J. Dairy Sci.* **2014**, *97*, 1047–1056. [CrossRef]
148. Ruiz-Cortés, Z.T.; Yoder, P.; Hanigan, M.D. Effects of Essential Amino Acid Deficiency on General Control Nonderepressible 2/Eukaryotic Initiation Factor 2 Signaling and Proteomic Changes in Primary Bovine Mammary Epithelial Cells. *Curr. Issues Mol. Biol.* **2022**, *44*, 1075–1086. [CrossRef]
149. Jiang, N.; Wang, Y.; Yu, Z.; Hu, L.; Liu, C.; Gao, X.; Zheng, S. WISP3 (CCN6) regulates milk protein synthesis and cell growth through mTOR signaling in dairy cow mammary epithelial cells. *DNA Cell Biol.* **2015**, *34*, 524–533. [CrossRef] [PubMed]
150. Tayebati, S.K.; Marucci, G.; Santinelli, C.; Buccioni, M.; Amenta, F. Choline-containing phospholipids: Structure-activity relationships versus therapeutic applications. *Curr. Med. Chem.* **2015**, *22*, 4328–4340. [CrossRef]
151. Glier, M.B.G.; Timothy, D.J.; Angela, M. Methyl nutrients, DNA methylation, and cardiovascular disease. *Mol. Nutr. Food Res.* **2014**, *58*, 172–182. [CrossRef]
152. Veth, M.J.D.; Artegoitia, V.M.; Campagna, S.R.; Lapierre, H.; Harte, F.; Girard, C.L. Choline absorption and evaluation of bioavailability markers when supplementing choline to lactating dairy cows. *J. Dairy Sci.* **2016**, *99*, 9732–9744. [CrossRef]
153. Cheng, R.; MacDonald, C.; Williams, C.; Meck, W. Prenatal choline supplementation alters the timing, emotion, and memory performance (TEMP) of adult male and female rats as indexed by differential reinforcement of low-rate schedule behavior. *Learn Mem.* **2008**, *15*, 153–162. [CrossRef]
154. Artegoitia, V.M.; Middleton, J.L.; Harte, F.M.; Campagna, S.R.; Veth, M.J. Choline and choline metabolite patterns and associations in blood and milk during lactation in dairy cows. *PLoS ONE* **2014**, *9*, e103412. [CrossRef]
155. Eklund, M.; Bauer, E.; Wamatu, J.; Mosenthin, R. Potential nutritional and physiological functions of betaine in livestock. *Nutr. Res. Rev.* **2005**, *18*, 31–48. [CrossRef]
156. Koc, H.; Mar, M.H.; Ranasinghe, A.; Swenberg, J.A.; Zeisel, S.H. Quantitation of choline and its metabolites in tissues and foods by liquid chromatography/electrospray ionization-isotope dilution mass spectrometry. *Anal. Chem.* **2002**, *74*, 4734–4740. [CrossRef]
157. Siljander, R.H.; Peuranen, S.; Tiihonen, K.; Virtanen, E.; Kettunen, H.; Alaviuhkola, T.; Simmins, P. Effect of equimolar dietary betaine and choline addition on performance, carcass quality, and physiological parameters of pigs. *Anim. Sci.* **2003**, *76*, 55–62. [CrossRef]
158. Zeisel, S. Choline, other methyl-donors and epigenetics. *Nutrients* **2017**, *9*, 445. [CrossRef]
159. Esposito, G.; Irons, P.C.; Webb, E.C.; Chapwanya, A. Interactions between negative energy balance, metabolic diseases, uterine health and immune response in transition dairy cows. *Anim. Reprod. Sci.* **2014**, *144*, 60–71. [CrossRef] [PubMed]
160. Itle, A.J.; Huzzey, J.M.; Weary, D.M.; VonKeyserlingk, M.A.G. Clinical ketosis and standing behavior in transition cows. *J. Dairy Sci.* **2015**, *98*, 128–134. [CrossRef] [PubMed]
161. Overton, T.R.; Waldron, M.R. Nutritional management of transition dairy cows: Strategies to optimize metabolic health. *J. Dairy Sci.* **2004**, *87* (Suppl. SE), 105–119. [CrossRef]
162. Pinotti, L.; Baldi, A.; Politis, I.; Rebucci, R.; Sangalli, L.; Dell'Orto, V. Rumen-protected choline administration to transition cows: Effects on milk production and vitamin E status. *J. Vet. Med. Ser. A* **2003**, *50*, 18–21. [CrossRef]
163. Castillo, C.; Hemandez, J.; Bravo, A.; Lopez, A.M.; Pereira, V.; Benedito, J.L. Oxidative status during late pregnancy and early lactation in dairy cows. *J. Vet.* **2005**, *169*, 286–292. [CrossRef]
164. LeBlanc, S.J.; Herdt, T.H.; Seymour, W.M.; Duffield, T.F.; Leslie, K.E. Pripartum serum vitamin E, retinol, and beta-carotene in dairy cattle and their associations with disease. *J. Dairy Sci.* **2004**, *87*, 609–619. [CrossRef]
165. Elek, P.; Gaál, T.; Husvéth, F. Influence of rumen-protected choline on liver composition and blood variables indicating energy balance in periparturient dairy cows. *Acta Vet. Hung.* **2013**, *61*, 59–70. [CrossRef]
166. Zenobi, M.G.; Scheffler, T.L.; Zuniga, J.E.; Poindexter, M.B.; Campagna, S.R.; Gonzalez, H.C.; Farmer, A.T.; Barton, B.A.; Santos, J.E.; Staples, C.R. Feeding increasing amounts of ruminally protected choline decreased fatty liver in nonlactating, pregnant Holstein cows in negative energy status. *J. Dairy Sci.* **2018**, *101*, 5902–5923. [CrossRef]

167. Lyu, Z. Effects of Dietary Rumen-Protected Choline Supplementation on Colostrum Yields, Quality, and Choline Metabolites from Dairy Cattle. *Highlights Sci. Eng. Technol.* **2022**, *14*, 266–269. [CrossRef]
168. Swartz, T.H.; Bradford, B.J.; Malysheva, O.; Caudill, M.A.; Mamedova, L.K.; Estes, K.A. Effects of dietary rumen-protected choline supplementation on colostrum yields, quality, and choline metabolites from dairy cattle. *JDS Commun.* **2022**, *3*, 296–300. [CrossRef]
169. Amrutkar, S.A.; Pawar, S.P.; Thakur, S.S.; Kewalramani, N.J.; Mahesh, M.S. Dietary Supplementation of Rumen-Protected Methionine, Lysine and Choline Improves Lactation Performance and Blood Metabolic Profile of Karan-Fries Cows. *Agric. Res.* **2015**, *4*, 396–404. [CrossRef]
170. White, H. Feeding for a Healthy Liver: The Role of Methionine and Choline in Transition Cows. *WCDS Adv. Dairy Technol.* **2020**, *32*, 87–92.
171. Abbasi, I.H.; Abbasi, F.; Soomro, R.N.; Abd El-Hack, M.E.; Abdel-Latif, M.A.; Li, W.; Hao, R.; Sun, F.; Bodinga, B.M.; Hayat, K.; et al. Considering choline as methionine precursor, lipoproteins transporter, hepatic promoter, and antioxidant agent in dairy cows. *AMB Express* **2017**, *7*, 214. [CrossRef] [PubMed]
172. Imhasly, S.; Bieli, C.; Naegeli, H.; Nyström, L.; Ruetten, M.; Gerspach, C. Blood plasma lipidome profile of dairy cows during the transition period. *BMC Vet. Res.* **2015**, *11*, 252. [CrossRef] [PubMed]
173. Cooke, R.F.; Del Rio, N.S.; Caraviello, D.Z.; Bertics, S.J.; Ramos, M.H.; Grummer, R.R. Supplemental choline for prevention and alleviation of fatty liver in dairy cattle. *J. Dairy Sci.* **2007**, *90*, 2413–2418. [CrossRef] [PubMed]
174. Lima, F.S.; Bisinotto, R.S.; Ribeiro, E.S.; Greco, L.F.; Ayres, H.; Favoreto, M.G.; Carvalho, M.R.; Galvão, K.N.; Santos, J.E. Effects of 1 or 2 treatments with prostaglandin F2α on subclinical endometritis and fertility in lactating dairy cows inseminated by timed artificial insemination. *J. Dairy Sci.* **2013**, *96*, 6480–6488. [CrossRef] [PubMed]
175. Baldi, A.; Pinotti, L. Choline metabolism in high-producing dairy cows: Metabolic and nutritional basis. *Can. J. Anim. Sci.* **2006**, *86*, 207–212. [CrossRef]
176. Zenobi, M.; Gardinal, R.; Zuniga, J.; Dias, A.; Nelson, C.; Driver, J.; Barton, B.; Santos, J.; Staples, C. Effects of supplementation with ruminally protected choline on performance of multiparous Holstein cows did not depend upon prepartum caloric intake. *J. Dairy Sci.* **2018**, *101*, 1088–1110. [CrossRef]
177. Hartwell, J.R.; Cecavqa, M.J.; Donkin, S.S. Impact of dietary rumen undegradable protein and rumen-protected choline on intake, peripartum liver triacylglyceride, plasma metabolites and milk production in transition dairy cows. *J. Dairy Sci.* **2000**, *83*, 2907–2917. [CrossRef]
178. Bakr, M.; Mohamed, S.H. Effect of rumen-protected choline supplementation on productive performance of lactating dairy cows. *Egypt. J. Anim. Prod.* **2020**, *57*, 113–120.
179. Bollatti, J.M.; Zenobi, M.G.; Barton, B.A.; Staples, C.R.; Santos, J.E. Responses to rumen-protected choline in transition cows do not depend on prepartum body condition. *J. Dairy Sci.* **2020**, *103*, 2272–2286. [CrossRef]
180. Elek, P.; Newbold, J.R.; Gaál, T.; Wagner, L.; Husveth, F. Effects of rumen-protected choline supplementation on milk production and choline supply of periparturient dairy cows. *Animal* **2008**, *2*, 1595–1601. [CrossRef]
181. Holdorf, H.T.; Kendall, S.J.; Ruh, K.E.; Caputo, M.J.; Combs, G.J.; Henisz, S.J.; Brown, W.E.; Bresolin, T.; Ferreira, R.E.; Dorea, J.R.; et al. Increasing the prepartum dose of rumen-protected choline: Effects on milk production and metabolism in high-producing Holstein dairy cows. *J. Dairy Sci.* **2023**, *106*, 5988–6004. [CrossRef]
182. Kumar, R.; Nayak, S.; Baghel, R.P.S.; Malapure, C.D.; Roy, B. Consequence of Prill Fat and Rumen Protected Choline supplementation on Milk yield and its makeup in Murrah Buffaloes. *Buffalo Bull.* **2019**, *38*, 49–56.
183. Mecionytė, I.; Palubinskas, G.; Anskienė, L.; Japertienė, R.; Juodžentytė, R.; Žilaitis, V. The effect of Supplementation of Rumen-Protected Choline on Reproductive and Productive Performances of Dairy Cows. *Animals* **2022**, *12*, 1807. [CrossRef] [PubMed]
184. Mohsen, M.K.; Gaafar, H.M.A.; Khalafalla, M.M.; Shitta, A.A.; Yousif, A.M. Effect of rumen protected choline supplementation on digestibility, rumen activity, and milk yield in lactating Friesian cows. *Slovak J. Anim. Sci.* **2011**, *44*, 13–20.
185. Morrison, E.I.; Reinhardt, H.; Leclerc, H.; DeVries, T.J.; LeBlanc, S.J. Effect of rumen-protected B vitamins and choline supplementation on health, production, and reproduction in transition dairy cows. *J. Dairy Sci.* **2018**, *101*, 9016–9027. [CrossRef]
186. Potts, S.; Scholte, C.; Moyes, K.; Erdman, R. Production responses to rumen-protected choline and methionine supplemented during the periparturient period differ for primi- and multiparous cows. *J. Dairy Sci.* **2020**, *103*, 6070–6086. [CrossRef] [PubMed]
187. Shahsavari, A.; D'Occhio, M.; Al Jassim, R. The role of rumen-protected choline in hepatic function and performance of transition dairy cows. *Br. J. Nutr.* **2016**, *116*, 35–44. [CrossRef] [PubMed]
188. Swartz, T.; Bradford, B.; Mamedova, L.; Estes, K. Effects of dietary rumen-protected choline supplementation to periparturient dairy cattle on inflammation, metabolism, and performance during an intramammary lipopolysaccharide challenge. *J. Dairy Sci.* **2023**, *in press*. [CrossRef]
189. Zahra, L.C.; Duffield, T.F.; Leslie, K.E.; Overton, T.R.; Putnam, D.; LeBlanc, S.J. Effects of rumen-protected choline and monensin on milk production and metabolism of periparturient cows. *J. Dairy Sci.* **2006**, *89*, 4808–4818. [CrossRef]
190. Zom, R.; van Baal, J.; Goselink, R.; Bakker, J.; de Veth, M.; van Vuuren, A. Effect of rumen-protected choline on performance, blood metabolites, and hepatic triacylglycerols of periparturient dairy cattle. *J. Dairy Sci.* **2011**, *94*, 4016–4027. [CrossRef]

191. Holdorf, H.T.; Brown, W.E.; Combs, G.J.; Henisz, S.J.; Kendall, S.J.; Caputo, M.J.; Ruh, K.E.; White, H.M. Increasing the Prepartum Dose of Rumen-Protected Choline: Effects of Maternal Choline Supplementation on Growth, Feed Efficiency, and Metabolism in Holstein and Holstein × Angus Calves. *J. Dairy Sci.* **2023**, *106*, 6005–6027. [CrossRef]
192. Sherlock, D.N.; Abdel-Hamied, E.; Bucktrout, R.; Liang, Y.; Miura, M.; Loor, J.J. Post-ruminal Choline Supply during Negative Nutrient Balance Alters Components of Hepatic mTOR Signaling and Plasma Amino Acids in Lactating Holstein Cows. *J. Dairy Sci.* **2023**, *in press*. [CrossRef]
193. Du, X.E.; Cui, Z.; Zhang, R.; Zhao, K.; Wang, L.; Yao, J.; Liu, S.; Cai, C.; Cao, Y. The Effects of Rumen-Protected Choline and Rumen-Protected Nicotinamide on Liver Transcriptomics in Periparturient Dairy Cows. *Metabolites* **2023**, *13*, 594. [CrossRef] [PubMed]

MDPI AG
Grosspeteranlage 5
4052 Basel
Switzerland
Tel.: +41 61 683 77 34

Metabolites Editorial Office
E-mail: metabolites@mdpi.com
www.mdpi.com/journal/metabolites

www.ingramcontent.com/pod-product-compliance
Lightning Source LLC
Chambersburg PA
CBHW062052220926
51850CB00010B/2748
9783725868803